INTRODUCTION TO
DATA TECHNOLOGIES

Chapman & Hall/CRC

Computer Science and Data Analysis Series

The interface between the computer and statistical sciences is increasing, as each discipline seeks to harness the power and resources of the other. This series aims to foster the integration between the computer sciences and statistical, numerical, and probabilistic methods by publishing a broad range of reference works, textbooks, and handbooks.

SERIES EDITORS

Proposals for the series should be sent directly to one of the series editors above, or submitted to:

Chapman & Hall/CRC
4th Floor, Albert House
1-4 Singer Street
London EC2A 4BQ
UK

Published Titles

Bayesian Artificial Intelligence
Kevin B. Korb and Ann E. Nicholson

Computational Statistics Handbook with MATLAB®, Second Edition
Wendy L. Martinez and Angel R. Martinez

Pattern Recognition Algorithms for Data Mining
Sankar K. Pal and Pabitra Mitra

Exploratory Data Analysis with MATLAB®
Wendy L. Martinez and Angel R. Martinez

Clustering for Data Mining: A Data Recovery Approach
Boris Mirkin

Correspondence Analysis and Data Coding with Java and R
Fionn Murtagh

Design and Modeling for Computer Experiments
Kai-Tai Fang, Runze Li, and Agus Sudjianto

Introduction to Data Technologies
Paul Murrell

Introduction to Machine Learning and Bioinformatics
Sushmita Mitra, Sujay Datta, Theodore Perkins, and George Michailidis

R Graphics
Paul Murrell

R Programming for Bioinformatics
Robert Gentleman

Semisupervised Learning for Computational Linguistics
Steven Abney

Statistical Computing with R
Maria L. Rizzo

Computer Science and Data Analysis Series

INTRODUCTION TO DATA TECHNOLOGIES

PAUL MURRELL

CRC Press is an imprint of the
Taylor & Francis Group, an **informa** business
A CHAPMAN & HALL BOOK

Chapman & Hall/CRC
Taylor & Francis Group
6000 Broken Sound Parkway NW, Suite 300
Boca Raton, FL 33487-2742

Printed in the United States of America on acid-free paper
10 9 8 7 6 5 4 3 2 1

International Standard Book Number-13: 978-1-4200-6517-6 (Hardcover)

Library of Congress Cataloging-in-Publication Data

Murrell, Paul.
Introduction to data technologies / Paul Murrell.
p. cm. -- (Chapman & Hall/CRC computer science and data analysis series)
Includes bibliographical references and index.
ISBN 978-1-4200-6517-6 (hardcover : alk. paper)
1. Computer science. 2. Electronic data processing. I. Title. II. Series.

QA76.M894 2009
004--dc22 2009001085

Visit the Taylor & Francis Web site at
http://www.taylorandfrancis.com

and the CRC Press Web site at
http://www.crcpress.com

to

Dominic, Matthew

and

Christina

Contents

List of Figures

List of Tables

Preface

Conducting research is a bit like parenting.

Raising a child involves a lot of cleaning and tidying, setting standards, and maintaining order, all of which goes completely unnoticed and for which the parent receives absolutely no credit.

Similarly, producing a bright, shiny result from the raw beginnings of a research project involves a lot of work that is almost never seen or acknowledged. Data sets never pop into existence in a fully mature and reliable state; they must be cleaned and massaged into an appropriate form. Just getting the data ready for analysis often represents a significant component of a research project.

Another thing that parenting and the "dirty jobs" of research have in common is that nobody gets taught how to do it. Parents just learn on the job and researchers typically have to do likewise when it comes to learning how to manage their data.

The aim of this book is to provide important information about how to work with research data, including ideas and techniques for performing the important behind-the-scenes tasks that take up so much time and effort, but typically receive little attention in formal education.

The focus of this book is on computational tools. The intention is to improve the awareness of what sorts of tasks can be achieved and to describe the correct approach to performing these tasks. There is also an emphasis on working with data technologies by typing computer code at a keyboard, rather than using a mouse to select menus and dialog boxes.

This book will not turn the reader into a web designer, or a database administrator, or a software engineer. However, this book contains information on how to publish information via the world wide web, how to access information stored in different formats, and how to write small programs to automate simple, repetitive tasks. A great deal of information on these topics already exists in books and on the internet; the value of this book is in collecting only the important subset of this information that is necessary to begin applying these technologies within a research setting.

Who should read this book?

This is an *introductory* computing book. It was originally developed as support for a second-year university course for statistics students and assumes no background in computing other than the ability to use a keyboard and mouse, and the ability to locate documents (or files) within folders (or directories). This means that it is suitable for anyone who is not confident about his or her computer literacy.

However, the book should also serve as a quick start on unfamiliar topics even for an experienced computer user, as long as the reader is not offended by over-simplified explanations of familiar material.

The book is also directed mainly at educating individual researchers. The tools and techniques, at the level they are described in this book, are of most use for the activities of a a single researcher or the activities of a small research team.

For people involved in managing larger projects, expert data management assistance is advisable. Nevertheless, a familiarity with the topics in this book will be very useful for communicating with experts and understanding the important issues.

In summary, this book is primarily aimed at research students and individual researchers with little computing experience, but it is hoped that it will also be of use to a broader audience.

Writing code

The icon below was captured from the desktop of a computer running Microsoft Windows XP.

Is this document a Microsoft Office Excel spreadsheet?

Many computer users would say that it is. After all, it has got the little Excel image on it and it even says Microsoft Office Excel right below the name of the file. And if we double-clicked on this file, Excel would start up and open the file.

However, this file is not an Excel spreadsheet. It is a plain text file in a Comma-Separated Values (CSV) format. In fact, the name of the file is not "final", but "final.csv". Excel can open this file, but so can thousands of other computer programs.

The computer protects us from this gritty detail by not showing the .csv suffix on the filename and it provides the convenience of automatically using Excel to open the file, rather than asking us what program to use.

Is this somehow a bad thing?

Yes, it is.

A computer user who only works with this sort of interface learns that this sort of file is only for use with Excel. The user becomes accustomed to the computer dictating what the user is able to do with a file.

It is important that users understand that we are able to dictate to the computer what should be done with a file. A CSV file can be viewed and modified using software such as Microsoft Notepad, but this may not occur to a user who is used to being told to use Excel.

Another example is that many computer users have been led to believe that the only way to view a web page is with Internet Explorer, when in fact there are many different web browsers *and* it is possible to access web pages using other kinds of software as well.

For the majority of computer users, interaction with a computer is limited to clicking on web page hyperlinks, selecting menus, and filling in dialog boxes. The problem with this approach to computing is that it gives the impression that the user is controlled by the computer. The computer interface places limits on what the user can do.

The truth is of course exactly the opposite. It is the computer user who has control and can tell the computer exactly what to do. The standard desktop PC is actually a "universal computing machine". It can do (almost) anything!

Learning to interact with a computer by writing computer code places users in their rightful position of power.

Computer code also has the huge advantage of providing an accurate record of the tasks that were performed. This serves both as a reminder of what was done and a recipe that allows others to replicate what was done.

For these reasons, this book focuses on computer languages as tools for data management.

Open standards and open source

This book almost exclusively describes technologies that are described by open standards or that are implemented in open source software, or both.

For a technology to be an open standard, it must be described by a public document that provides enough information so that anyone can write software to work with technology. In addition, the description must not be subject to patents or other restrictions of use. Ideally, the document is published and maintained by an international, non-profit organisation. In practice, the important consequence is that the technology is not bound to a single software product.

This is in contrast to proprietary technologies, where the definitive description of the technology is not made available and is only officially supported by a single software product.

Open source software is software for which the source code is publicly available. This makes it possible, through scrutiny of the source code if necessary, to understand how a software product works. It also means that, if necessary, the behavior of the software can be modified. In practice, the important consequence is that the software is not bound to a single software developer.

This is in contrast to proprietary software, where the software is only available from a single developer, the software is a "black-box", and changes, including corrections of errors, can only be made by that software developer.

The obvious advantage of using open standards and open source software is that the reader need not purchase any expensive proprietary software in order to benefit from the information in this book, but that is not the primary reason for this choice.

The main reason for selecting open standards and open source software is that this is the only way to ensure that we know, or can find out, where our data are on the computer and what happens to our data when we manipulate the data with software, and it is the only way to guarantee that we can have free access to our data now and in the future.

The significance of these points is demonstrated by the growing list of governments and public institutions that are switching to open standards and open source software for storing and working with information. In particular, for the storage of public records, it does not make sense to lock the information up in a format that cannot be accessed except by proprietary software. Similarly, for the dissemination and reproducibility of research, it makes sense to fully disclose a complete description of how an analysis was conducted in addition to publishing the research results.

How this book is organized

This book is designed to be accessible and practical, with an emphasis on useful, applicable information. To this end, each topic is introduced via one or more case studies, which helps to motivate the need for the relevant ideas and tools. Practical examples are used to demonstrate the most important points and there is a deliberate avoidance of minute detail. Separate reference chapters then provide a more structured and detailed description for a particular technology, which is more useful for finding specific information once the big picture has been obtained. These reference chapters are still not exhaustive, so pointers to further reading are also provided.

The main topics are organized into four core chapters, with supporting reference chapters, as described below.

Chapter 2: Writing Computer Code

This chapter discusses how to write computer code, using the HyperText Markup Language, HTML, as a concrete example. A number of important ideas and terminology are introduced for working with any computer language, and it includes guidelines and advice on the practical aspects of how to write computer code in a disciplined way. HTML provides a way to produce documents that can be viewed in a web browser and published on the world wide web.

Chapters 3 and 4 provide support in the form of reference material for HTML and Cascading Style Sheets.

Chapter 5: Data Storage

This chapter covers a variety of data storage topics, starting with a range of different file formats, which includes a brief discussion of how data values are stored in computer memory, moving on to a discussion of the eXtensible Markup Language, XML, and ending up with the structure and design issues of relational databases.

Chapter 6 provides reference material for XML and the Document Type Definition language.

Chapter 7: Data Queries

This chapter focuses on accessing data, with a major focus on extracting data from a relational database using the Structured Query

Language, SQL. There is also a brief mention of the XPath language for accessing data in XML documents.

Chapter 8 provides reference material for SQL, including additional uses of SQL for creating and modifying relational databases.

Chapter 9: Data Processing

This chapter is by far the largest. It covers a number of tools and techniques for searching, sorting, and tabulating data, plus several ways to manipulate data to change the data into new forms. This chapter introduces some very basic programming concepts and introduces the R language for statistical computing.

Chapter 10 provides reference material for R and Chapter 11 provides reference material for regular expressions, which is a language for processing text data.

Chapter 12 provides a brief wrap-up of the main ideas in the book.

There is an overall progression through the book from writing simple computer code with straightforward computer languages to more complex tasks with more sophisticated languages. The core chapters also build on each other to some extent. For example, Chapter 9 assumes that the reader has a good understanding of data storage formats and is comfortable writing computer code. Furthermore, examples and case studies are carried over between different chapters in an attempt to illustrate how the different technologies need to be combined over the lifetime of a data set. There are also occasional "flashbacks" to a previous topic to make explicit connections between similar ideas that reoccur in different settings. In this way, the book is set up to be read in order from start to finish.

However, every effort has been made to ensure that individual chapters can be read on their own. Where necessary, figures are reproduced and descriptions are repeated so that it is not necessary to jump back and forth within the book in order to acquire a complete understanding of a particular section.

Much of the information in this book will require practice in order to gain a full understanding. The reader is encouraged to make use of the exercises on the book's web site.

The web site

There is an accompanying web site for this book at:

`http://www.stat.auckland.ac.nz/~paul/ItDT/`

This site includes complete PDF and HTML versions of the book, code and data sets used in the case studies, and a suite of exercises.

Software

The minimum software requirements for making use of the information in this book are the following open source products:

Mozilla's Firefox web browser
`http://www.mozilla.com/en-US/products/firefox/`

The SQLite database engine
`http://www.sqlite.org/`

The R language and environment
`http://www.r-project.org/`

About the license

Acknowledgements

Many people have helped in the creation of this book, in a variety of ways.

First of all, thanks to those who have provided useful comments, suggestions, and corrections on the manuscript, particularly the anonymous reviewers. I am also indebted to those who have donated data sets for use in the case studies within this book and for use in the exercises that are available from the book web site. Because I expect these lists to grow over time, the names of these contributors are all posted on the book web site.

Another vote of thanks must go to the authors of the images that I have used in various places throughout the book. In most cases, I have had no direct contact with these authors because they have generously released their work under permissive licenses that allow their work to be shared and reused. The full list of attributions for these works is given on page 401.

I would also like to acknowledge Kurt Hornik and Achim Zeileis from the Computational Statistics group of the Department of Statistics and Mathematics at the Wirtschaftsuniversität Wien in Vienna for providing me with a productive environment during the latter half of 2007 to work on this book. Thanks also to my host institution, the University of Auckland.

Last, and most, thank you, Ju.

Paul Murrell
The University of Auckland
New Zealand

This manuscript was generated with LaTeX, gawk, make, latex2html, GIMP, ghostscript, ImageMagick, Sweave, R, SQLite, and GNU/Linux.

1

Introduction

1.1 Case study: Point Nemo

The Pacific Ocean is the largest body of water on Earth and accounts for almost a third of the Earth's surface area.

The Live Access Server is one of many services provided by the National Aeronautics and Space Administration (NASA) for gaining access to its enormous repositories of atmospheric and astronomical data. The Live Access Server[1] provides access to atmospheric data from NASA's fleet of Earth-observing satellites, data that consist of coarsely gridded measurements of major atmospheric variables, such as ozone, cloud cover, pressure, and temperature. NASA provides a web site that allows researchers to select variables of interest, and geographic and temporal ranges, and then to download or view the relevant data (see Figure 1.1). Using this service, we can attempt to answer questions about atmospheric and weather conditions in different parts of the world.

The Pacific Pole of Inaccessibility is a location in the Southern Pacific Ocean that is recognized as one of the most remote locations on Earth. Also known as Point Nemo, it is the point on the ocean that is farthest from any land mass. Its counterpart, the Eurasian Pole of Inaccessibility, in northern China, is the location on land that is farthest from any ocean.

These two geographical extremes—one in the southern hemisphere, over 2,500 km from the nearest land, and one in the northern hemisphere, over 2,500 km from the nearest ocean—are usually only of interest either to intrepid explorers or conspiracy theorists (a remote location is the perfect

[1] `http://mynasadata.larc.nasa.gov/LASintro.html`

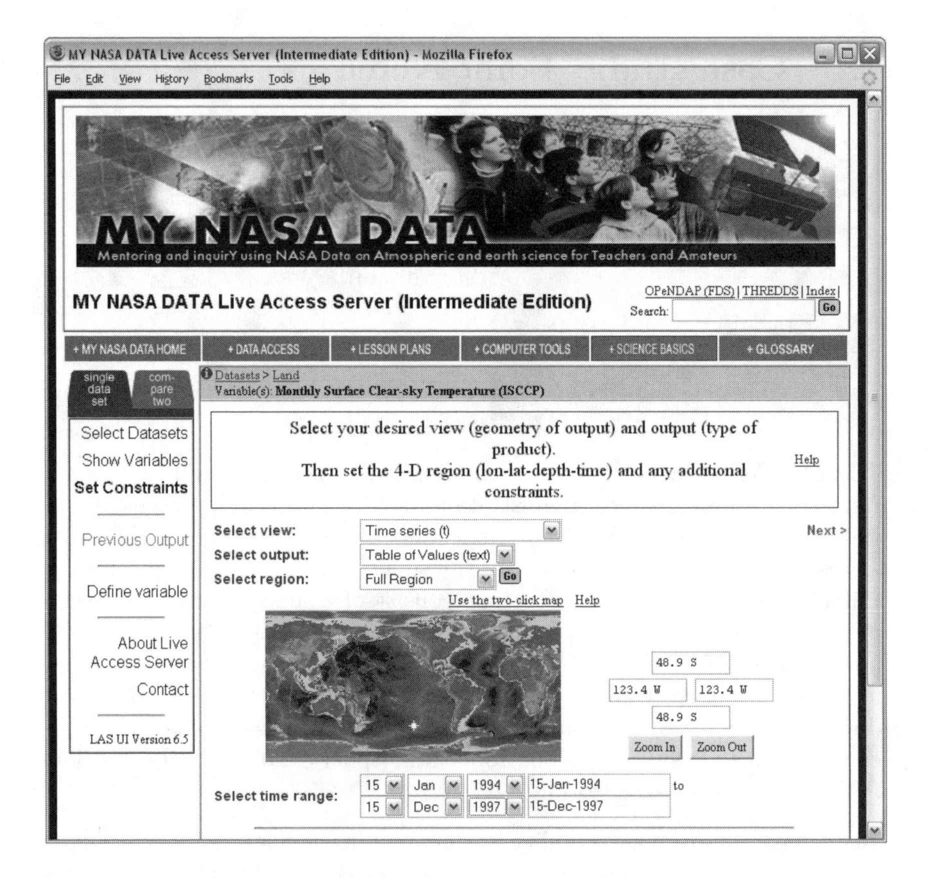

Figure 1.1: NASA's Live Access Server web site. On the map, the Pacific Pole of Inaccessibility is marked with a white plus sign.

place to hide an important secret!). However, our interest will be to investigate the differences in weather conditions between these interesting geographical extremes by using NASA's Live Access Server.

To make our task a little more manageable, for now we will restrict our attention to a comparison of the surface temperatures at each of the Poles of Inaccessibility. To be precise, we will look at monthly average temperatures at these locations from January 1994 to December 1997.

In a book on data analysis, we would assume that the data are already in a form that can be conveniently loaded into statistical software, and the emphasis would be on how to analyze these data. However, that is not the focus of this book. Here, we are interested in all of the steps that must be taken *before* the data can be conveniently loaded into statistical software.

As anyone who has worked with data knows, it often takes more time and effort to get the data ready than it takes to perform the data analysis. And yet there are many more books on how to analyze data than there are on how to prepare data for analysis. This book aims to redress that balance.

In our example, the main data collection has already occurred; the data are measurements made by instruments on NASA satellites. However, we still need to collect the data from NASA's Live Access Server. We will do this initially by entering the appropriate parameters on the Live Access Server web site. Figure 1.2 shows the first few lines of data that the Live Access Server returns for the surface temperature at Point Nemo.

The first thing we should always do with a new data set is take a look at the raw data. Viewing the raw data is an important first step in becoming familiar with the data set. We should never automatically assume that the data are reliable or correct. We should always check with our own eyes. In this case, we are already in for a bit of a shock.

Anyone who expects temperatures to be in degrees Celsius will find values like 278.9 something of a shock. Even if we expect temperatures on the Fahrenheit scale, 278.9 is hotter than the average summer's day.

The problem of course is that these are scientific measurements, so the scale being used is Kelvin; the temperature scale where zero really means zero. 278.9 K is 5.8°C or 42°F, which is a cool, but entirely believable, surface temperature value. When planning a visit to Point Nemo, it would be a good idea to pack a sweater.

Looking at the raw data, we also see a lot of other information besides the surface temperatures. There are longitude and latitude values, dates, and a description of the variable that has been measured, including the units

```
            VARIABLE : Mean TS from clear sky composite (kelvin)
            FILENAME : ISCCPMonthly_avg.nc
            FILEPATH : /usr/local/fer_data/data/
            SUBSET   : 48 points (TIME)
            LONGITUDE: 123.8W(-123.8)
            LATITUDE : 48.8S
                      123.8W
                       23
16-JAN-1994 00 /  1:  278.9
16-FEB-1994 00 /  2:  280.0
16-MAR-1994 00 /  3:  278.9
16-APR-1994 00 /  4:  278.9
16-MAY-1994 00 /  5:  277.8
16-JUN-1994 00 /  6:  276.1
...
```

Figure 1.2: The first few lines of plain text output from the Live Access Server for the surface temperature at Point Nemo.

of measurement. This **metadata** is very important because it provides us with a proper understanding of the data set. For example, the metadata makes it clear that the temperature values are on the Kelvin scale. The metadata also tells us that the longitude and latitude values, 123.8 W and 48.8 S, are not exactly what we asked for. It turns out that the values provided by NASA in this data set have been averaged over a large area, so this is as good as we are going to get.

Before we go forward, we should take a step back and acknowledge the fact that we are able to read the data at all. This is a benefit of the **storage format** that the data are in; in this case, it is a **plain text** format. If the data had been in a more sophisticated **binary** format, we would need something more specialized than a common web browser to be able to view our data. In Chapter 5 we will spend a lot of time looking at the advantages and disadvantages of different data storage formats.

Having had a look at the raw data, the next step in familiarizing ourselves with the data set should be to look at some numerical summaries and plots. The Live Access Server does not provide numerical summaries and, although it will produce some basic plots, we will need a bit more flexibility. Thus, we will save the data to our own computer and load it into a statistical software package.

The first step is to save the data. The Live Access Server will provide an **ASCII file** for us, or we can just copy-and-paste the data into a **text editor** and save it from there. Again, we should appreciate the fact that this step is quite straightforward and is likely to work no matter what sort of computer or operating system we are using. This is another feature of having data in a plain text format.

Now we need to get the data into our statistical software. At this point, we encounter one of the disadvantages of a plain text format. Although we, as human readers, can see that the surface temperature values start on the ninth line and are the last value on each row (see Figure 1.2), there is no way that statistical software can figure this out on its own. We will have to describe the format of the data set to our statistical software.

In order to read the data into statistical software, we need to be able to express the following information: "skip the first 8 lines"; and "on each row, the values are separated by whitespace (one or more spaces or tabs)"; and "on each row, the date is the first value and the temperature is the last value (ignore the other three values)". Here is one way to do this for the statistical software package R (Chapter 9 has much more to say about working with data in R):

```
read.table("PointNemo.txt", skip=8,
           colClasses=c("character",
                        "NULL", "NULL", "NULL",
                        "numeric"),
           col.names=c("date", "", "", "", "temp"))
```

This solution may appear complex, especially for anyone not experienced in writing computer code. Partly that is because this is complex information that we need to communicate to the computer and writing code is the best or even the only way to express that sort of information. However, the complexity of writing computer code gains us many benefits. For example, having written this piece of code to load in the data for the Pacific Pole of Inaccessibility, we can use it again, with only a change in the name of the file, to read in the data for the Eurasian Pole of Inaccessibility. That would look like this:

```
read.table("Eurasia.txt", skip=8,
           colClasses=c("character",
                        "NULL", "NULL", "NULL",
                        "numeric"),
           col.names=c("date", "", "", "", "temp"))
```

Imagine if we wanted to load in temperature data in this format from several hundred other locations around the world. Loading in such volumes of data would now be trivial and fast using code like this; performing such a task by selecting menus and filling in dialog boxes hundreds of times does not bear thinking about.

Going back a step, if we wanted to download the data for hundreds of locations around the world, would we want to fill in the Live Access Server web form hundreds of times? Most people would not. Here again, we can write code to make the task faster, more accurate, and more bearable.

As well as the web interface to the Live Access Server, it is also possible to make requests by **writing code** to communicate with the Live Access Server. Here is the code to ask for the temperature data from the Pacific Pole of Inaccessibility:

```
lasget.pl \
    -x -123.4 -y -48.9 -t 1994-Jan-1:2001-Sep-30 \
    -f txt \
    http://mynasadata.larc.nasa.gov/las-bin/LASserver.pl \
    ISCCPMonthly_avg_nc ts
```

Again, that may appear complex, and there is a "start-up" cost involved in learning how to write such code. However, this is the only sane method to obtain large amounts of data from the Live Access Server. Chapters 7 and 9 look at extracting data sets from complex systems and automating tasks that would otherwise be tedious or impossible if performed by hand.

Writing code, as we have seen above, is the only accurate method of communicating even mildly complex ideas to a computer, and even for very simple ideas, writing code is the most efficient method of communication. In this book, we will always communicate with the computer by **writing code**. In Chapter 2 we will discuss the basic ideas of how to write computer code properly and we will encounter a number of different computer languages throughout the remainder of the book.

At this point, we have the tools to access the Point Nemo data in a form that is convenient for conducting the data analysis, but, because this is not a book on data analysis, this is where we stop. The important points for our purposes are how the data are stored, accessed, and processed. These are the topics that will be expanded upon and discussed at length throughout the remainder of this book.

Summary

This book is concerned with the issues and technologies involved with the storage and handling of data sets.

We will focus on the ways in which these technologies can help us to perform tasks more efficiently and more accurately.

We will emphasize the appropriate use of these technologies, in particular, the importance of performing tasks by writing computer code.

2

Writing Computer Code

There are two aims for this chapter: learning how to write computer code and learning a computer language to write code in.

First, we need to learn how to write computer code. Several of the computer technologies that we encounter will involve writing computer code in a particular computer language, so it is essential that we learn from the start how to produce computer code in the right way.

Learning how to write code could be very dull if we only discussed code writing in abstract concepts, so the second aim of this chapter is to learn a computer language, with which to demonstrate good code writing.

The language that we will learn in this chapter is the Hypertext Markup Language, HTML.

As most people know, HTML is the language that describes web pages on the world wide web. In the world wide web of today, web pages are used by almost everyone for almost any purpose—sharing videos, online sales, banking, advertising, the list goes on and on—and web pages consist of much more than just plain HTML.

However, the original motivation for HTML was a simple, intuitive platform that would allow researchers from all over the world to publish and share their ideas, data, and results.

The primary inventor of HTML was Tim Berners-Lee, the father of several fundamental technologies underlying the world wide web. His original work was driven by a number of important requirements: it should be simple to create web pages; web pages should work on any computer hardware and software; and web pages should be able to link to each other, so that information can be related and cross-referenced, but remain distributed around the world. It was also an early intention to provide a technology that could be processed and understood by computers as well as viewed by human eyes.

This history and these requirements make HTML an appropriate starting point for learning about computer languages for the management of research data.

HTML is simple, so it is a nice easy way to start writing computer code. HTML began as an open technology and remains an open standard, which is precisely the sort of technology we are most interested in. We require only a web browser to view web pages and these are widely available on any modern desktop computer. We will also, later in this chapter, observe a resonance with the idea that HTML documents simply point to each other rather than require a copy of every piece of information on every computer. In Chapter 9, we will return to the importance of being able to automatically process HTML.

Finally, although HTML is now used for all sorts of commercial and private purposes, it still remains an important technology for publishing and sharing research output.

The aim of this chapter is to elucidate the process of writing, checking, and running computer code and to provide some guidelines for carrying out these tasks effectively and efficiently.

How this chapter is organized

This chapter begins with an example of a simple web page and gives a quick introduction to how HTML computer code relates to the web pages that we see in a web browser such as Firefox. The main point is to demonstrate that computer code can be used to control what the computer does.

In Section 2.2, we will emphasize the idea that computer languages have strict rules that must be followed. Computer code must be exactly correct before it can be used to control the computer. The specific rules for HTML code are introduced in this section.

Section 2.3 addresses the issue of what computer code *means*. What instructions do we have to use in order to make the computer do what we want? This section looks in a bit more detail at some examples of HTML code and shows how each piece of code relates to a specific feature of the resulting web page.

Section 2.4 looks at the practical aspects of writing computer code. We will discuss the software that should be used for writing code and emphasize the importance of writing *tidy* and well-organized code.

Sections 2.5 and 2.6 look at getting our computer code to work properly. Sections 2.5 focuses on the task of checking that computer code is correct—that it follows the rules of the computer language—and Section 2.6 focuses on how to *run* computer code. We will look at some tools for checking that HTML code is correct and we will briefly discuss the software that produces web pages from HTML code. We will also discuss the process of fixing problems when they occur.

Section 2.7 introduces some more ideas about how to organize code and work efficiently. The Cascading Style Sheets language is introduced to provide some simple demonstrations.

2.1 Case study: Point Nemo (continued)

Figure 2.1 shows a simple web page that contains a very brief statistical report on the Poles of Inaccessibility data from Chapter 1. The report consists of paragraphs of text and an image (a plot), with different formatting applied to various parts of the text. The report heading is bold and larger than the other text, and the table of numerical summaries is displayed with a monospace font. Part of the heading and part of the supplementary material at the bottom of the page are italicized and also act as hyperlinks; a mouse click on the underlined text in the heading will navigate to the Wikipedia page on the Poles of Inaccessibility and a click on the underlined text at the bottom of the page will navigate to NASA's Live Access Server home page.

The entire web page in Figure 2.1 is described using HTML. The web page consists of a simple text file called `inaccessibility.html` that contains HTML code. The HTML code is shown in Figure 2.2.

We do not need to worry about the details of this HTML code yet. What is important is to notice that all of this code is just text. Some of it is the text that makes up the content of the report, and other parts are special HTML **keywords** that "mark up" the content and describe the nature and purpose of each part of the report. The keywords can be distinguished as the text that is surrounded by angled brackets. For example, the heading at the top of the page consists of two keywords, `<h3>` and `</h3>` surrounding the actual text of the heading (see lines 7 to 12).

The HTML code is just text, so the overall document is just a text file. We will explore the differences between different sorts of files in more detail in Chapter 5. For now, it is just important to note that HTML code, and any other computer code that we write, should be written using software that creates plain text files. For example, do *not* write computer code using a word processor, such as Microsoft Word. We will return to this point in Section 2.4.1.

Another feature of the code in Figure 2.2 is that it has a clear structure. For example, for every **start tag**, there is an **end tag**; the `<head>` start tag on line 3 is matched by the `</head>` closing tag on line 5. This rigid

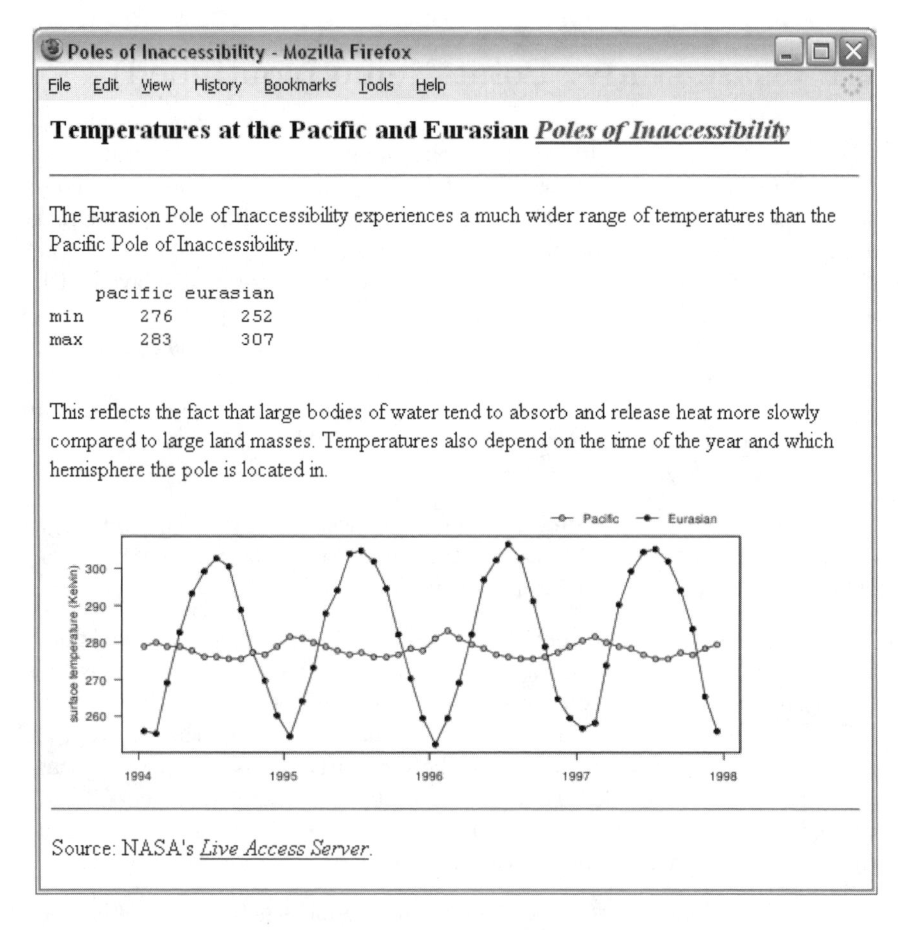

Figure 2.1: A simple web page that displays information about the surface temperature data for the Pacific and Eurasian Poles of Inaccessibility (viewed with the Firefox web browser on Windows XP).

structure is an important feature of computer languages and reflects the fact that there are strict rules to follow for any computer language. It is vital that we observe this structure, but the discipline required will help us to be accurate, clear, and logical in how we think about tasks and in how we communicate our instructions to the computer. As we will see later in this chapter, it is also important that this structure is reflected in the layout of our code (as has been done in Figure 2.2).

In this chapter we will learn the basics of HTML, with a focus on how the code itself is written and with an emphasis on the correct way to write computer code.

2.2 Syntax

The first thing we need to learn about a computer language is the correct **syntax** for the language.

Syntax is essentially the punctuation and grammar rules for a computer language. Certain characters and words have special meanings and must appear in a particular order for the computer code to make any sense.

A simple example from line 3 in Figure 2.2 is the piece of HTML code `<head>`. The `<` character is special in HTML because it indicates the start of a keyword in HTML. The `>` character is also special; it marks the end of the keyword. Also, this particular keyword, `<head>`, must appear at the very start of a piece of HTML code.

These rules and special meanings for specific characters and words make up the syntax for a computer language.

Computers are extremely fussy when it comes to syntax, so computer code will have to have all syntax rules perfectly correct before the code will work.

The next section describes the basic syntax rules for the HTML language.

2.2.1 HTML syntax

HTML has a very simple syntax.

HTML code consists of two basic components: **elements**, which are special HTML keywords, and **content**, which is just normal everyday text.

There are a few elements that have to go in every HTML document—Figure 2.3 shows the smallest possible HTML document—and then it is up to the

```
 1 <!DOCTYPE HTML PUBLIC "-//W3C//DTD HTML 4.01 Transitional//EN">
 2 <html>
 3   <head>
 4     <title>Poles of Inaccessibility</title>
 5   </head>
 6   <body>
 7     <h3>
 8     Temperatures at the Pacific and Eurasian
 9     <a href="http://wikipedia.org/wiki/Pole_of_inaccessibility"
10        style="font-style: italic">
11     Poles of Inaccessibility</a>
12     </h3>
13
14     <hr>
15     <p>
16     The Eurasion Pole of Inaccessibility experiences a much
17     wider range of temperatures than the Pacific Pole of
18     Inaccessibility.
19     </p>
20
21     <pre>
22     pacific eurasian
23 min     276      252
24 max     283      307
25     </pre>
26
27     <p>
28     This reflects the fact that large bodies of water tend to
29     absorb and release heat more slowly compared to large
30     land masses.
31     Temperatures also depend on the time of the year and which
32     hemisphere the pole is located in.
33     </p>
34
35     <img src="poleplot.png">
36
37     <hr>
38     <p>
39     Source:  NASA's
40     <a href="http://mynasadata.larc.nasa.gov/LASintro.html"
41        style="font-style: italic">
42     Live Access Server</a>.
43     </p>
44   </body>
45 </html>
```

Figure 2.2: The file `inaccessibility.html`, which contains the HTML code behind the web page in Figure 2.1. The line numbers (in grey) are just for reference.

```
<!DOCTYPE HTML PUBLIC "-//W3C//DTD HTML 4.01 Transitional//EN">
<html>
    <head>
        <title></title>
    </head>
    <body>
    </body>
</html>
```

Figure 2.3: A minimal HTML document. This is the basic code that must appear in any HTML document. The main content of the web page is described by adding further HTML elements within the `body` element.

author to decide on the main contents of the web page.

An HTML element consists of a **start tag**, an **end tag** and some content in between.

As an example, we will look closely at the `title` element from line 4 of Figure 2.2.

```
<title>Poles of Inaccessibility</title>
```

This code is broken down into its separate components below, with one important component labeled and highlighted (underlined) on each row.

start tag: `<title>Poles of Inaccessibility</title>`
content: `<title>Poles of Inaccessibility</title>`
end tag: `<title>Poles of Inaccessibility</title>`

The greater-than and less-than signs have a special meaning in HTML code; they mark the start and end of HTML tags. All of the characters with a special meaning in this `title` element are highlighted below.

special characters: `<title>Poles of Inaccessibility</title>`

Some HTML elements may be **empty**, which means that they only consist of a start tag (no end tag and no content). An example is the `img` (short for "image") element from Figure 2.2, which inserts the plot in the web page.

```
<img src="poleplot.png">
```

The entire `img` element consists of this single tag.

There is a fixed set of valid HTML elements and only those elements can be used within HTML code. We will encounter several important elements in this chapter and a more comprehensive list is provided in Chapter 3.

Attributes

HTML elements can have one or more **attributes**, which provide more information about the element. An attribute consists of the attribute name, an equals sign, and the attribute value, which is surrounded by double-quotes. Attributes only appear in the start tag of an element. We have just seen an example in the `img` element above. The `img` element has an attribute called `src` that describes the location of a file containing the picture to be drawn on the web page. In the example above, the attribute is `src="poleplot.png"`.

The components of this HTML element are shown below.

```
      HTML tag:  <img src="poleplot.png">
  element name:  <img src="poleplot.png">
     attribute:  <img src="poleplot.png">
attribute name:  <img src="poleplot.png">
attribute value: <img src="poleplot.png">
```

Again, some of the characters in this code are part of the HTML syntax. These special characters are highlighted below.

```
special characters:  <img src="poleplot.png">
```

Many attributes are optional, and if they are not specified, a default value is provided.

Element order

HTML tags must be ordered properly. All elements must nest cleanly and some elements are only allowed inside specific other elements. For example, a `title` element can only be used inside a `head` element, and the `title` element must start and end within the `head` element. The following HTML code is invalid because the `title` element does not finish within the `head` element.

```
<head>
    <title>
    Poles of Inaccessibility
</head>
    </title>
```

To be correct, the `title` element must start and end within the `head` element, as in the code below.

```
<head>
    <title>
    Poles of Inaccessibility
    </title>
</head>
```

Finally, there are a few elements that *must* occur in an HTML document: there must be a `DOCTYPE` declaration, which states what computer language we are using; there must be a single `html` element, with a single `head` element and a single `body` element inside; and the `head` element must contain a single `title` element. Figure 2.3 shows a minimal HTML document.

HTML is defined by a standard, so there is a single, public specification of HTML syntax. Unfortunately, as is often the case, there are several different *versions* of HTML, each with its own standard, so it is necessary to specify exactly which version of HTML we are working with. We will focus on HTML version 4.01 in this book. This is specified in the `DOCTYPE` declaration used in all examples.

These are the basic syntax rules of HTML. With these rules, we can write *correct* HTML code. In Section 2.3 we will look at the next step, which is what the code will do when we give it to the computer to run.

2.2.2 Escape sequences

As we have seen in HTML, certain words or characters have a special meaning within the language. These are sometimes called **keywords** or **reserved words** to indicate that they are reserved by the language for special use and cannot be used for their normal natural-language purpose.

This means that some words or characters can never be used for their normal, natural-language meaning when writing in a formal computer language and a special code must be used instead.

For example, the `<` character marks the start of a tag in HTML, so this cannot be used for its normal meaning of "less than".

If we need to have a less-than sign within the content of an HTML element, we have to type `<` instead. This is an example of what is called an **escape sequence**.

Another special character in HTML is the greater-than sign, `>`. To produce one of these in the content of an HTML element, we must type `>`.

In HTML, there are several escape sequences of this form that all start with an ampersand, `&`. This means of course that the ampersand is itself a special character, with its own escape sequence, `&`. A larger list of special characters and escape sequences in HTML is given in Section 3.1.2.

We will meet this idea of escape sequences again in the other computer languages that we encounter later in the book.

Recap

Computer code is just text, but with certain characters or words having special meanings.

The punctuation and grammar rules of a computer language are called the syntax of the language.

Computer code must have all syntax rules correct before it can be expected to work properly.

An escape sequence is a way of getting the normal meaning for a character or word that has a special meaning in the language.

HTML consists of elements, which consist of a start tag and an end tag, with content in between.

HTML elements may also have attributes within the start tag.

2.3 Semantics

When we write code in a computer language, we call the meaning of the code—what the computer will do when the code is run—the **semantics** of the code. Computer code has no defined semantics until it has a correct syntax, so we should always check that our code is free of errors before worrying about whether it does what we want.

Computer languages tend to be very precise, so as long as we get the syntax right, there should be a clear meaning for the code. This is important because it means that we can expect our code to produce the same result on different computers and even with different software.

2.3.1 HTML semantics

In HTML, tags are used to mark up and identify different parts of a document. When the HTML is viewed in a web browser, different parts of the document are displayed in different ways. For example, headings are typically drawn larger and bolder than normal text and paragraphs are typeset so that the text fills the space available, but with visual gaps from one paragraph to the next.

The HTML 4.01 specification defines a fixed set of valid HTML elements and describes the meaning of each of those elements in terms of how they should be used to create a web page.

In this section, we will use the simple HTML page shown at the start of this chapter to demonstrate some of the basic HTML elements. Chapter 3 provides a larger list.

Figure 2.1 shows what the web page looks like and 2.2 shows the underlying HTML code.

The main part of the HTML code in Figure 2.2 is contained within the `body` element (lines 6 to 44). This is the content of the web page—the information that will be displayed by the web browser.

In brief, this web page consists of: an `h3` element to produce a heading; several `p` elements that produce paragraphs of text; two `hr` elements that produce horizontal lines; an `img` element that generates the plot; and a `pre` element that produces the table of numerical summaries.

The first element we encounter within the `body` is an `h3` element (lines 7 to 12). The contents of this element provide a title for the page, which is indicated by drawing the relevant text bigger and bolder than normal text. There are several such heading elements in HTML, from `h1` to `h6`, where the number indicates the heading "level", with `1` being the top level (biggest and boldest) and `6` the lowermost level. Note that this element does two things: it describes the *structure* of the information in the web page and it controls the *appearance* for the information—how the text should be displayed. The structure of the document is what we should focus on; we will discuss the appearance of the web page in more depth in Section 2.7.1.

The next element in our code is an hr (horizontal rule) element (line 14). This produces a horizontal line across the page, just below the heading.

Next is a p (paragraph) element (lines 15 to 19). The p tags indicate that the contents should be arranged as a paragraph of text. The important thing to notice about this element is that the web browser decides where to break the lines of text; the line breaks and whitespace within the HTML code are ignored.

The next element in our code is a pre (preformatted text) element (lines 21 to 25). This element is used to display text exactly as it is entered, using a monospace font, with all spaces faithfully reproduced. This is in contrast to the previous p element, where the layout of the text in the HTML code bore no relation to the layout within the web page.

The next element is another paragraph of text (lines 27 to 33), and then we have an img (image) element (line 35). One difference with this element is that it has an attribute. The src attribute specifies the location of the file for the image.

At the bottom of the page we have another hr element (line 37), followed by a final paragraph (p element) of text (lines 38 to 43).

Part of the text in the final paragraph is also a hyperlink. The a (anchor) element around the text "Live Access Server" (lines 9 and 11) means that this text is highlighted and underlined. The href attribute specifies that when the text is clicked, the browser will navigate to the home page of NASA's Live Access Server.

These are some of the simple HTML elements that can be used to create web pages. Learning the meaning of these HTML keywords allows us to produce HTML code that is not only correct, but also produces the web page that we want.

There are many more HTML elements that can be used to create a variety of different effects and Chapter 3 describes a few more of these.

Recap

The meaning of keywords in a computer language is called the semantics of the language.

A web page is described by placing HTML tags around the content of the page.

2.4 Writing code

Up to this point, we have only looked at HTML code that has already been written. We will now turn our attention to writing *new* HTML code.

There are three important steps: we must learn how to write the code in the first place; we must be able to check the syntax of the code; and we must learn how to run the code to produce the desired end result (in the case of HTML, a web page).

For each step, we will discuss what software we need to do the job as well as provide guidelines and advice on the right way to perform each task.

In this section, we look at the task of *writing* computer code.

2.4.1 Text editors

The act of writing code is itself dependent on computer tools. We use software to record and manage our keystrokes in an effective manner. This section discusses what sort of tool should be used to write computer code effectively.

An important feature of computer code is that it is just plain text. There are many software packages that allow us to enter text, but some are more appropriate than others.

For many people, the most obvious software program for entering text is a word processor, such as Microsoft Word or Open Office Writer. These programs are *not* a good choice for editing computer code. A word processor is a good program for making text look pretty with lots of fancy formatting and wonderful fonts. However, these are not things that we want to do with our raw computer code.

The programs that we use to run our code expect to encounter only plain text, so we must use software that creates only text documents, which means we must use a **text editor**.

2.4.2 Important features of a text editor

For many people, the most obvious program for creating a document that only contains text is Microsoft Notepad. This program has the nice feature that it saves a file as pure text, but its usefulness ends there.

When we write computer code, a good choice of text editor can make us much more accurate and efficient. The following facilities are particularly useful for writing computer code:

automatic indenting
: As we will see in Section 2.4.3, it is important to arrange code in a neat fashion. A text editor that helps to indent code (place empty space at the start of a line) makes this easier and faster.

parenthesis matching
: Many computer languages use special symbols, e.g., `{` and `}`, to mark the beginning and end of blocks of code. Some text editors provide feedback on such matching pairs, which makes it easier to write code correctly.

syntax highlighting
: All computer languages have special **keywords** that have a special meaning for the language. Many text editors automatically color such keywords, which makes it easier to read code and easier to spot simple mistakes.

line numbering
: Some text editors automatically number each line of computer code (and in some cases each column or character as well) and this makes navigation within the code much easier. This is particularly important when trying to find errors in the code (see Section 2.5).

In the absence of everything else, Notepad is better than using a word processor. However, many useful (and free) text editors exist that do a much better job. Some examples are Crimson Editor on Windows[1] and Kate on Linux.[2]

Figure 2.4 demonstrates some of these ideas by showing the same code in Notepad and Crimson Editor.

2.4.3 Layout of code

There are two important audiences to consider when writing computer code. The obvious one is the computer; it is vitally important that the computer understands what we are trying to tell it to do. This is mostly a matter of getting the syntax of our code right.

[1]`http://www.crimsoneditor.com/`
[2]`http://kate-editor.org/`

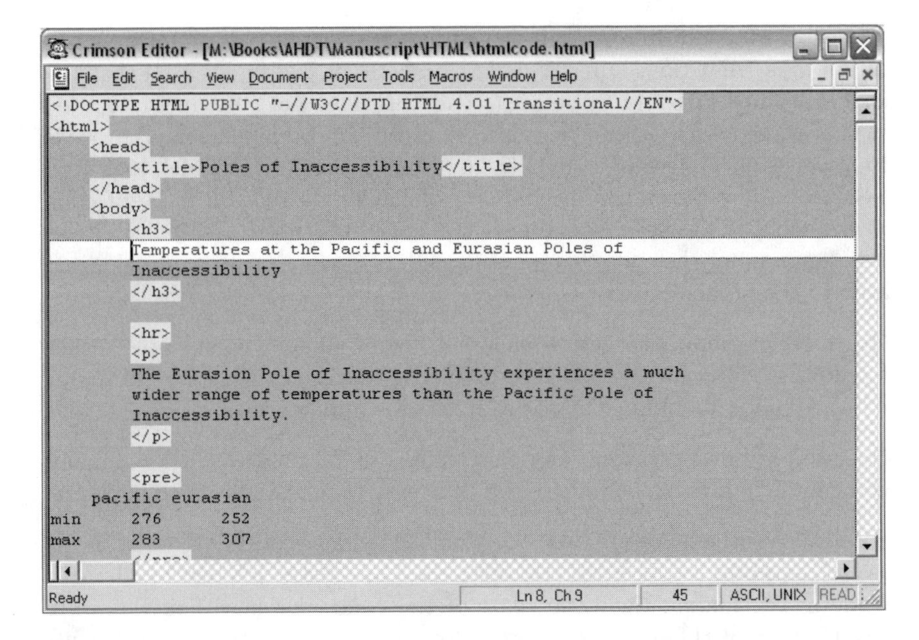

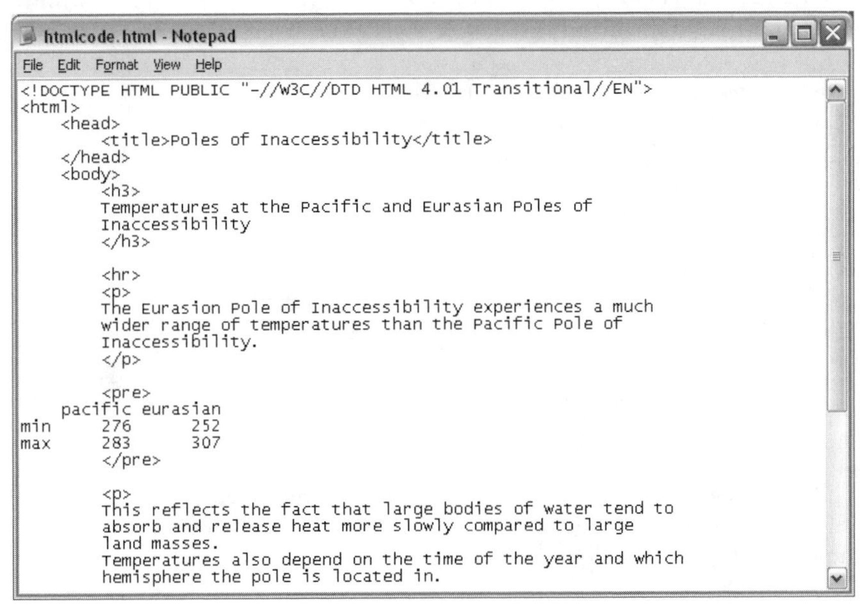

Figure 2.4: The HTML code from Figure 2.2 viewed in Crimson Editor (top) and Microsoft Notepad (bottom). Crimson Editor provides assistance for writing computer code by providing syntax highlighting (the HTML keywords are highlighted) and by providing information about which row and column the cursor is on (in the status bar at the bottom of the window).

The other audience for code consists of humans. While it is important that code works (that the computer understands it), it is also essential that the code is comprehensible to people. And this does not just apply to code that is shared with others, because the most important person who needs to understand a piece of code is the original author of the code! It is very easy to underestimate the probability of having to reuse a piece of code weeks, months, or even years after it was initially written, and in such cases it is common for the code to appear much less obvious on a second viewing, even to the original author.

Other people may also get to view a piece of code. For example, other researchers will want to see our code so that they know what we did to our data. All code should be treated as if it is for public consumption.

One simple but important way that code can be improved for a human audience is to format the code so that it is easy to read and easy to navigate.

For example, the following two code chunks are identical HTML code, as far as the computer is concerned. However, they are vastly different to a human reader. Try finding the "title" part of the code. Even without knowing anything about HTML, this is a ridiculously easy task in the second layout, and annoyingly difficult in the first.

```
<html><head><title>A Minimal HTML
Document</title></head><body>
The content goes here!</body>
```

```
<html>
    <head>
        <title>A Minimal HTML Document</title>
    </head>
    <body>
        The content goes here!
    </body>
```

This demonstrates the basic idea behind laying out code. The changes are entirely cosmetic, but they are extremely effective. It also demonstrates one important layout technique: indenting.

2.4.4 Indenting code

The idea of indenting code is to expose the *structure* of the code. What this means will vary between computer languages, but in the case of HTML code, a simple rule is to indent the contents of an element.

The following code provides a simple example, where a `title` element is the content of a `head` element. The `title` element is indented (shifted to the right) with respect to the `head` element.

```
<head>
    <title>A Minimal HTML Document</title>
</head>
```

The amount of indenting is a personal choice. The examples here have used 4 spaces, but 2 spaces or even 8 spaces are also common. Whatever indentation is chosen, it is essential that the indenting rule is applied consistently, especially when more than one person might modify the same piece of code.

Exposing structure of code by indenting is important because it makes it easy for someone reading the code to navigate within the code. It is easy to identify different parts of the code, which makes it easier to see what the code is doing.

Another useful result of indenting is that it provides a basic check on the correctness of code. Look again at the simple HTML code example. Does anything look wrong?

```
<html>
    <head>
        <title>A Minimal HTML Document</title>
    </head>
    <body>
        The content goes here!
    </body>
```

Even without knowing anything about HTML, the lack of symmetry in the layout suggests that there is something missing at the bottom of this piece of code. In this case, indenting has alerted us to the fact that there is no end `</html>` tag.

2.4.5 Long lines of code

Another situation where indenting should be applied is when a line of computer code becomes very long. It is a bad idea to have a single line of code that is wider than the screen on which the code is being viewed (so that we have to scroll across the window to see all of the code). When this happens, the code should be split across several lines (most computer languages do not notice the difference). Here is an example of a line of HTML code that is too long.

```
<img src="poleplot.png" alt="A plot of temperatures over time">
```

Here is the code again, split across several lines. It is important that the subsequent lines of code are indented so that they are visually grouped with the first line.

```
<img src="poleplot.png"
     alt="A plot of temperatures over time">
```

In the case of a long HTML element, a reasonable approach is to left-align the start of all attributes within the same tag (as shown above).

2.4.6 Whitespace

Whitespace refers to empty gaps in computer code. Like indenting, whitespace is useful for making code easy for humans to read, but it has no effect on the semantics of the code. Wouldyouwriteinyournativelanguagewithoutputtingspacesbetweenthewords?

Indenting is a form of whitespace that always appears at the start of a line, but whitespace is effective *within* and *between* lines of code as well. For example, the following code is too dense and therefore is difficult to read.

```
<table border="1"width="100%"bgcolor="#CCCCCC">
```

This modification of the code, with extra spaces, is much easier on the eye.

```
<table border="1" width="100%" bgcolor="#CCCCCC">
```

Figure 2.5 shows two code chunks that demonstrate the usefulness of blank lines between code blocks to help expose the structure, particularly in large pieces of code.

Again, exactly when to use spaces or blank lines depends on personal style.

2.4.7 Documenting code

In Section 2.5.3, we discuss the importance of being able to *read* documentation about a computer language. In this section, we consider the task of

```
<!DOCTYPE HTML PUBLIC "-//W3C//DTD HTML 4.01//EN">
<html>
    <head>
        <title>Poles of Inaccessibility</title>
    </head>
    <body>
        <h3>
        Temperatures at the Pacific and Eurasian
        Poles of Inaccessibility
        </h3>
        <hr>
        <p>
        The Eurasion Pole of Inaccessibility experiences
        a much wider range of temperatures than the
        Pacific Pole of Inaccessibility.
        </p>
        <pre>
    pacific eurasian
min     276     252
max     283     307
        </pre>
        <p>
        This reflects the fact that large bodies of
        water tend to absorb and release heat more
        slowly compared to large land masses.
        Temperatures also depend on the time of the year
        and which hemisphere the pole is located in.
        </p>
        <img src="poleplot.png">
        <hr>
        <p style="font-style: italic">
        Source:  NASA's
        <a href="http://mynasadata.larc.nasa.gov/">
        Live Access Server</a>.
        </p>
    </body>
</html>
```

```
<!DOCTYPE HTML PUBLIC "-//W3C//DTD HTML 4.01//EN">
<html>
    <head>
        <title>Poles of Inaccessibility</title>
    </head>
    <body>
        <h3>
        Temperatures at the Pacific and Eurasian
        Poles of Inaccessibility
        </h3>

        <hr>
        <p>
        The Eurasion Pole of Inaccessibility experiences
        a much wider range of temperatures than the
        Pacific Pole of Inaccessibility.
        </p>

        <pre>
    pacific eurasian
min     276     252
max     283     307
        </pre>

        <p>
        This reflects the fact that large bodies of
        water tend to absorb and release heat more
        slowly compared to large land masses.
        Temperatures also depend on the time of the year
        and which hemisphere the pole is located in.
        </p>

        <img src="poleplot.png">

        <hr>
        <p style="font-style: italic">
        Source:  NASA's
        <a href="http://mynasadata.larc.nasa.gov/">
        Live Access Server</a>.
        </p>
    </body>
</html>
```

Figure 2.5: These two code chunks contain exactly the same code; all that differs is the use of several blank lines (whitespace) in the code on the right, which help to expose the structure of the code for a human reader.

writing documentation for our own code.

As with the layout of code, the purpose of documentation is to communicate. The obvious target of this communication is other people, so that they know what we did. A less obvious, but no less important, target is the code author. It is essential that when we return to a task days, weeks, or even months after we first performed the task, we are able to pick up the task again, and pick it up quickly.

Most of what we will have to say about documentation will apply to writing **comments**—messages written in plain language, embedded in the code, and which the computer ignores.

2.4.8 HTML comments

Here is how to include a comment within HTML code.

```
<!-- This is a comment -->
```

Anything between the start `<!--` and end `-->`, including HTML tags, is completely ignored by the computer. It is only there to edify a human reader.

Having no comments in code is generally a bad idea, and it is usually the case that people do not add enough comments to their code. However, it can also be a problem if there are too many comments. If there are too many comments, it can become a burden to ensure that the comments are all correct if the code is ever modified. It can even be argued that too many comments make it hard to see the actual code!

Comments should not just be a repetition of the code. Good uses of comments include: providing a conceptual summary of a block of code; explaining a particularly complicated piece of code; and explaining arbitrary constant values.

Recap

Computer code should be written using a text editor.

Code should be written tidily so that it is acceptable for a human audience. Code should be indented, lines should not be too long, and there should be plenty of whitespace.

Code should include comments, which are ignored by the computer but explain the purpose of the code to a human reader.

2.5 Checking code

Knowing how to write the correct syntax for a computer language is not a guarantee that we *will* write the correct syntax for a particular piece of code. One way to check whether we have our syntax correct is to stare at it and try to see any errors. However, in this book, such a tedious, manual, and error-prone approach is discouraged because the computer is so much better at this sort of task.

In general, we will enlist the help of computer software to check that the syntax of our code is correct.

In the case of HTML code, there are many types of software that can check the syntax. Some web browsers provide the ability to check HTML syntax, but in general, just opening the HTML document in a browser is *not* a good way to check syntax.

The software that we will use to demonstrate HTML syntax checking is a piece of software called HTML Tidy.

2.5.1 Checking HTML code

HTML Tidy is a program for checking the syntax of HTML code. It can be downloaded from Source Forge[3] or it can be used via one of the online services provided by the World Wide Web Consortium (W3C).[4]

In order to demonstrate the use of HTML Tidy, we will check the syntax of the following HTML, which contains one deliberate mistake. This code has been saved in a file called `broken.html`.

For simple use of HTML Tidy, the only thing we need to know is the name of the HTML document and where that file is located. For example, the online services provide a button to select the location of an HTML file. To run HTML Tidy locally, the following command would be entered in a command window or terminal.

```
tidy broken.html
```

HTML Tidy checks the syntax of HTML code, reports any problems that it finds, and produces a suggestion of what the correct code should look like.

Figure 2.7 shows part of the output from running HTML Tidy on the simple HTML code in Figure 2.6.

[3] `http://tidy.sourceforge.net/`

[4] `http://cgi.w3.org/cgi-bin/tidy`, `http://validator.w3.org/`

```
1 <!DOCTYPE HTML PUBLIC "-//W3C//DTD HTML 4.01 Transitional//EN">
2 <html>
3     <head>
4         <title>A Minimal HTML Document
5     </head>
6     <body>
7     </body>
8 </html>
```

Figure 2.6: An HTML document that contains one deliberate mistake on line 4 (missing <> tag). The line numbers (in grey) are just for reference.

```
Parsing "broken.html"
line 5 column 5 - Warning: missing </title> before </head>

Info: Doctype given is "-//W3C//DTD HTML 4.01 Transitional//EN"
Info: Document content looks like HTML 4.01 Transitional
1 warning, 0 errors were found!
```

Figure 2.7: Part of the output from running HTML Tidy on the HTML code in Figure 2.6.

An important skill to develop for writing computer code is the ability to decipher warning and error messages that the computer displays. In this case, there is one error message.

2.5.2 Reading error information

The error (or warning) information provided by computer software is often very terse and technical. Reading error messages is a skill that improves with experience, and it is important to seek out any piece of useful information in a message. Even if the message as a whole does not make sense, if the message can only point us to the correct area within the code, our mistake may become obvious.

In general, when the software checking our code encounters a series of errors, it is possible for the software to become confused. This can lead to more errors being reported than actually exist. It is *always* a good idea to tackle the first error first, and it is usually a good idea to recheck code after fixing each error. Fixing the first error will sometimes eliminate or at least modify subsequent error messages.

The error from HTML Tidy in Figure 2.7 is this:

```
line 5 column 5 - Warning: missing </title> before </head>
```

To an experienced eye, the problem is clear, but this sort of message can be quite opaque for people who are new to writing computer code. A good first step is to make use of the information that is supplied about *where* the problem has occurred. In this case, we need to find the fifth character on line 5 of our code.

The line of HTML code in question is shown below.

```
    </head>
```

Column 5 on this line is the `<` at the start of the closing `head` tag.

Taken in isolation, it is hard to see what is wrong with this code. However, error messages typically occur only once the computer is convinced that something has gone wrong. In many cases, the error will actually be in the code somewhere *in front of* the exact location reported in the error message. It is usually a good idea to look in the general area specified by the error message, particularly on the lines immediately preceding the error.

Here is the location of the error message in a slightly broader context.

```
        <title>A Minimal HTML Document
    </head>
    <body>
```

The error message mentions both `title` and `head`, so we might guess that we are dealing with these elements. In this case, this case just confirms that we are looking at the right place in our code.

The message is complaining that the `</title>` tag is missing and that the tag should appear before the `</head>` tag. From the code, we can see that we have started a `title` element with a `<title>` start tag, but we have failed to complete the element; there is no `</title>` end tag.

In this case, the solution is simple; we just need to add the missing tag and *check the code with HTML Tidy again*, and everything will be fine.

Unfortunately, not all syntax errors can be resolved as easily as this. When the error is not as obvious, we may have to extract what information we can from the error message and then make use of another important skill: reading the **documentation** for computer code.

2.5.3 Reading documentation

The nice thing about learning a computer language is that the rules of grammar are usually quite simple, there are usually very few of them, and they are usually very consistent.

Unfortunately, computer languages are similar to natural languages in terms of vocabulary. The time-consuming part of learning a computer language involves learning all of the special words in the language and their meanings.

What makes this task worse is the fact that the reference material for computer languages, much like the error messages, can be terse and technical. As for reading error messages, practice and experience are the only known cures.

This book provides reference chapters for each of the computer languages that we encounter. Chapter 3 provides a reference for HTML.

These reference chapters are shorter and simpler than the official language documentation, so they should provide a good starting point for finding out a few more details about a language. When this still does not provide the answer, there are pointers to more thorough documentation at the end of each reference chapter.

Recap

Computer code should be checked for correctness of syntax before it can be expected to run.

Understanding computer error messages and understanding the documentation for a computer language are important skills in themselves that take practice and experience to master.

2.6 Running code

We have now discussed how to write code and how to check that the syntax of the code is correct. The final step is to *run* the code and have the computer produce the result that we want.

As with syntax checking, we need to use software to run the code that we have written.

In the case of HTML, there are many software programs that will run the code, but the most common type is a web browser, such as Internet Explorer or Firefox.

2.6.1 Running HTML code

All that we need to do to run our HTML code is to open the file containing our code with a web browser. The file does *not* have to be on another computer on the internet. In most browsers, there is a File menu with an option to open an HTML file on the local computer. We can then see whether the code has produced the result that we want.

Web browsers tend to be very lenient with HTML syntax. If there is a syntax error in HTML code, most web browsers try to figure out what the code should do (rather than reporting an error). Unfortunately, this can lead to problems where two different web browsers will produce different results from exactly the same code.

Another problem arises because most web browsers do not completely implement the HTML standards. This means that some HTML code will not run correctly on some browsers.

The solution to these problems, for this book, has two parts: we will not use a browser to check our HTML syntax (we will use HTML Tidy instead; see Section 2.5), and we will use a single browser (Firefox[5]) to define what a piece of HTML code should do. Furthermore, because we will only be using a simple subset of the HTML language, the chance of encountering ambiguous behavior is small.

If we run HTML code and the result is what we want, we are finished. However, more often than not, the result is *not* what we want. The next section looks at how to resolve problems in our code.

2.6.2 Debugging code

When we have code that has correct syntax and runs, but does not behave correctly, we say that there is a **bug** in our code. The process of fixing our code so that it does what we want it to is called **debugging** the code.

It is often the case that debugging code takes much longer than writing the code in the first place, so it is an important skill to acquire.

The source of common bugs varies enormously with different computer lan-

[5] `http://www.mozilla.com/en-US/products/firefox/`

guages, but there are some common steps we should take when fixing any sort of code:

Do not blame the computer:
There are two possible sources of problems: our code is wrong or the computer (or software used to run our code) is wrong. It will almost always be the case that our code is wrong. If we are completely convinced that our code is correct and the computer is wrong, we should go home, have a sleep, and come back the next day. The problem in our code will usually then become apparent.

Recheck the syntax:
Whenever a change is made to the code, check the syntax again before trying to run the code again. If the syntax is wrong, there is no hope that the code will run correctly.

Develop code incrementally:
Do not try to write an entire web page at once. Write a small piece of code and get that to work, then add more code and get that to work. When things stop working, it will be obvious which bit of code is broken.

Do one thing at a time:
Do not make more than one change to the code at a time. If several changes are made, even if the problem is cured, it will be difficult to determine which change made the difference. A common problem is introducing new problems as part of a fix for an old problem. Make one change to the code and see if that corrects the behavior. If it does not, then revert that change before trying something else.

Read the documentation:
For all of the computer languages that we will deal with in this book, there are official documents plus many tutorials and online forums that contain information and examples for writing code. Find them and read them.

Ask for help:
In addition to the copious manuals and tutorials on the web, there are many forums for asking questions about computer languages. The friendliness of theses forums varies and it is important to read the documentation before taking this step.

Chapter 3 provides some basic information about common HTML elements and Section 3.3 provides some good starting points for detailed documentation about HTML.

We will discuss specific debugging tools for some languages as we meet them.

Recap

Computer code should be tested to ensure that it not only works but also that it produces the right result.

If code does not produce the right result, the code should be modified in small, disciplined, and deliberate stages to approach the right result.

2.7 The DRY principle

One of the purposes of this book is to introduce and explain various technologies for working with data. We have already met one such technology, HTML, for producing reports on the world wide web.

Another purpose of this book is to promote the correct approach, or "best practice", for using these technologies. An example of this is the emphasis on writing code using computer languages rather than learning to use dialog boxes and menus in a software application.

In this section, we will look at another example of best practice, called the **DRY principle**,[6] which has important implications for how we manage the code that we write.

DRY stands for **Don't Repeat Yourself** and the principle is that there should only ever be *one copy* of any important piece of information.

The reason for this principle is that one copy is much easier to maintain than multiple copies; if the information needs to be changed, there is only one place to change it. In this way the principle promotes efficiency. Furthermore, if we lapse and allow several copies of a piece of information, then it is possible for the copies to diverge or for one copy to get out of date. From this perspective, having only one copy improves our accuracy.

To understand the DRY principle, consider what happens when we move to a new address. One of the many inconveniences of changing addresses involves letting everyone know our new address. We have to alert schools, banks, insurance companies, doctors, friends, etc. The DRY principle suggests that we should have only one copy of our address stored somewhere (e.g., at the

[6]"The Pragmatic Programmer", Andy Hunt and Dave Thomas (1999), Addison-Wesley.

post office) and everyone else should refer to that address. That way, if we change addresses, we only have to tell the post office the new address and everyone will see the change. In the current situation, where there are multiple copies of our address, it is easy for us to forget to update one of the copies when we change address. For example, we might forget to tell the bank, so all our bank correspondence will be sent to the wrong address!

The DRY principle will be very important when we discuss the storage of data (Chapter 5), but it can also be applied to computer code that we write. In the next section, we will look at one example of applying the DRY principle to writing computer code.

2.7.1 Cascading Style Sheets

Cascading Style Sheets (CSS) is a language that is used to describe how to display information. It is commonly used with HTML to control the appearance of a web page. In fact, the preferred way to produce a web page is to use HTML to indicate the *structure* of the information and CSS to specify the *appearance*.

We will give a brief introduction to CSS and then go on to show how the proper use of CSS demonstrates the DRY principle.

CSS syntax

We have actually seen two examples of CSS already in the HTML code from the very start of this chapter. The CSS code occurs on lines 10 and 41 of Figure 2.2; these lines, plus their surrounding context, are reproduced below.

```
<a href="http://wikipedia.org/wiki/Pole_of_inaccessibility"
   style="font-style: italic">
Poles of Inaccessibility</a>
```

```
<a href="http://mynasadata.larc.nasa.gov/LASintro.html"
   style="font-style: italic">
Live Access Server</a>.
```

In both cases, the CSS code occurs within the start tag of an HTML anchor (`a`) element. The anchor element has an attribute called `style`, and the value of that attribute is a CSS **property**.

```
font-style: italic
```

This property has the name `font-style` and the value `italic`; the effect in both cases is to make the text of the hyperlink *italic*.

One component of the CSS language is this idea of properties. The appearance of an element is specified by setting one or more CSS properties for that element. There are a number of different properties to control things like the font used for text, colors, and borders and shading.

One way to use CSS is to add CSS properties as attributes of individual HTML elements, as in the above example. However, this has the disadvantage of requiring us to add a CSS property to every HTML element that we want to control the appearance of.

Figure 2.8 shows another way to use CSS properties; this is a small modification of the HTML code from Figure 2.2. The result in a web browser is exactly the same; all that has changed is how the CSS code is provided.

The important differences in Figure 2.8 are that there is a `style` element in the `head` of the HTML code (lines 5 to 7) and the anchor elements (lines 12 and 13) no longer have a `style` attribute.

The new `style` element in the `head` of the HTML code is reproduced below.

```
<style type="text/css">
  a { font-style: italic; }
</style>
```

One important point is that this CSS code no longer just consists of a CSS property. This CSS code is a complete CSS **rule**, which consists of a **selector**, plus one or more properties.

The components of this CSS rule are shown below.

selector:	`a { font-style: italic; }`
open bracket:	`a { font-style: italic; }`
property name:	`a { font-style: italic; }`
colon:	`a { font-style: italic; }`
property value:	`a { font-style: italic; }`
semi-colon:	`a { font-style: italic; }`
close bracket:	`a { font-style: italic; }`

The brackets, the colon, and the semi-colon will be the same for any CSS

```
 1 <!DOCTYPE HTML PUBLIC "-//W3C//DTD HTML 4.01 Transitional//EN">
 2 <html>
 3   <head>
 4     <title>Poles of Inaccessibility</title>
 5     <style type="text/css">
 6       a { font-style: italic; }
 7     </style>
 8   </head>
 9   <body>
10     <h3>
11     Temperatures at the Pacific and Eurasian
12     <a href="http://wikipedia.org/wiki/Pole_of_inaccessibility">
13     Poles of Inaccessibility</a>
14     </h3>
15     <hr>
16     <p>
17     The Eurasion Pole of Inaccessibility experiences a much
18     wider range of temperatures than the Pacific Pole of
19     Inaccessibility.
20     </p>
21     <pre>
22     pacific eurasian
23 min     276      252
24 max     283      307
25     </pre>
26     <p>
27     This reflects the fact that large bodies of water tend to
28     absorb and release heat more slowly compared to large
29     land masses.
30     Temperatures also depend on the time of the year and which
31     hemisphere the pole is located in.
32     </p>
33     <img src="poleplot.png">
34     <hr>
35     <p>
36     Source:  NASA's
37     <a href="http://mynasadata.larc.nasa.gov/LASintro.html">
38     Live Access Server</a>.
39     </p>
40   </body>
41 </html>
```

Figure 2.8: The file `inaccessibilitystyle.html`. This HTML code will produce the same web page as Figure 2.1 when viewed in a web browser. This code is similar to the code in Figure 2.2 except that the CSS code is within a `style` element in the `head` element, rather than being within a `style` *attribute* within the `anchor` element.

rule.

The selector part of the CSS rule specifies the elements within the HTML document that the properties part of the rule is applied to.

In this case, the selector is just `a`, which means that this rule applies to `all` anchor elements within the HTML document.

This is one demonstration of how CSS supports the DRY principle. Instead of adding CSS code in a `style` attribute to every element within an HTML document, we can just add CSS code to a single `style` *element* at the top of the HTML document and the rules within that CSS code will be applied to all relevant elements within the document.

This is good, but we can do even better.

A third way to use CSS code is to create a separate text file containing just the CSS code and *refer* to that file from the HTML document.

Figure 2.9 shows a modification of the code in Figure 2.8 that makes use of this approach.

The important difference this time is that there is a `link` element in the `head` element (lines 5 and 6) rather than a `style` element. This new `link` element is reproduced below. Most of this element is standard for referencing a file of CSS code; the only piece that is specific to this example is the `href` attribute, which specifies the name of the file that contains the CSS code.

```
<link rel="stylesheet" href="reportstyle.css"
      type="text/css">
```

This link element tells a web browser to load the CSS code from the file `reportstyle.css` and apply the rules in that file. The contents of that file are shown in Figure 2.10.

A CSS file can contain one or more CSS rules. In this case, there is just one rule, which applies to anchor elements and says that the text for hyperlinks should be italic.

The advantage of having a separate file of CSS code like this is that now we can apply the CSS rule not just to all anchor elements in the HTML document `inaccessibility.html`, but to all anchor elements in any other HTML document as well. For example, if we want this style for all statistical reports that we write, then we can simply include a `link` element that refers to `reportstyle.css` in the HTML document for each report. This is a

```
 1 <!DOCTYPE HTML PUBLIC "-//W3C//DTD HTML 4.01 Transitional//EN">
 2 <html>
 3   <head>
 4     <title>Poles of Inaccessibility</title>
 5     <link rel="stylesheet" href="reportstyle.css"
 6           type="text/css">
 7   </head>
 8   <body>
 9     <h3>
10     Temperatures at the Pacific and Eurasian
11     <a href="http://wikipedia.org/wiki/Pole_of_inaccessibility">
12     Poles of Inaccessibility</a>
13     </h3>
14     <hr>
15     <p>
16     The Eurasion Pole of Inaccessibility experiences a much
17     wider range of temperatures than the Pacific Pole of
18     Inaccessibility.
19     </p>
20     <pre>
21     pacific eurasian
22 min     276      252
23 max     283      307
24     </pre>
25     <p>
26     This reflects the fact that large bodies of water tend to
27     absorb and release heat more slowly compared to large
28     land masses.
29     Temperatures also depend on the time of the year and which
30     hemisphere the pole is located in.
31     </p>
32     <img src="poleplot.png">
33     <hr>
34     <p>
35     Source:  NASA's
36     <a href="http://mynasadata.larc.nasa.gov/LASintro.html">
37     Live Access Server</a>.
38     </p>
39   </body>
40 </html>
```

Figure 2.9: The file `inaccessibilitycss.html`. This HTML code will produce the same web page as Figure 2.1 when viewed in a web browser. This code is similar to the code in Figure 2.8 except that the CSS code is within a separate text file called `reportstyle.css` and there is a `link` element (rather than a `style` element) in the `head` element that refers to the file of CSS code.

```
1 a {
2     font-style: italic;
3 }
```

Figure 2.10: The file `reportstyle.css`, which contains CSS code to control the appearance of anchor elements within HTML documents. The line numbers (in grey) are just for reference.

further example of the DRY principle as it applies to computer code; we have one copy of code that can be reused in many different places.

Other CSS rules

As mentioned previously, there are many other CSS properties for controlling other aspects of the appearance of elements within a web page. Some of these are described in Chapter 4.

Chapter 4 also describes some more advanced ideas about the *selectors* within CSS rules. For example, there are ways to specify a selector so that a CSS rule only applies to some anchor elements within an HTML document, rather than *all* anchor elements within the document.

Recap

The DRY principle states that there should be only one copy of important information.

Cascading style sheets control the appearance of HTML documents.

By having separate HTML and CSS files, it is easy and efficient to have several different HTML documents with the same general appearance.

2.8 Further reading

Code Complete
by Steve McConnell
2nd edition (2004) Microsoft Press.
Exhaustive discussion of ways of writing good computer code. Includes languages and advanced topics way beyond the scope of this book.

Summary

Writing computer code should be performed with a text editor to produce a plain text file.

Code should first be checked for correct syntax (spelling and grammar).

Code that has correct syntax can then be run to determine whether it performs as intended.

Code should be written for human consumption as well as for correctness.

Comments should be included in code and the code should be arranged neatly so that the structure of the code is obvious to human eyes.

HTML is a simple language for describing the structure of the content of web pages. It is a useful cross-platform format for producing reports.

CSS is a language for controlling the appearance of the content of web pages.

The separation of code for a web page into HTML and CSS helps to avoid duplication of code (an example of the DRY principle in action).

HTML and CSS code can be run in any web browser.

3 HTML Reference

HTML is a computer language used to create web pages. HTML code can be run by opening the file containing the code with any web browser.

The information in this chapter describes HTML 4.01, which is a W3C Recommendation.

3.1 HTML syntax

HTML code consists of HTML **elements**.

An element consists of a start **tag**, followed by the element **content**, followed by an end tag. A start tag is of the form <*elementName*> and an end tag is of the form </*elementName*>. The example code below shows a `title` element; the start tag is `<title>`, the end tag is `</title>`, and the content is the text: `Poles of Inaccessibility`.

```
<title>Poles of Inaccessibility</title>
```

```
start tag:  <title>Poles of Inaccessibility</title>
  content:  <title>Poles of Inaccessibility</title>
  end tag:  <title>Poles of Inaccessibility</title>
```

Some elements are **empty**, which means that they consist of only a start tag (no content and no end tag). The following code shows an `hr` element, which is an example of an empty element.

```
<hr>
```

An element may have one or more **attributes**. Attributes appear in the start tag and are of the form *attributeName*="*attributeValue*". The code below shows the start tag for an `img` element, with an attribute called `src`. The value of the attribute in this example is `"poleplot.png"`.

```
<img src="poleplot.png">
```

```
<!DOCTYPE HTML PUBLIC "-//W3C//DTD HTML 4.01 Transitional//EN">
<html>
    <head>
        <title></title>
    </head>
    <body>
    </body>
</html>
```

Figure 3.1: A minimal HTML document. This is the basic code that must appear in any HTML document. The main content of the web page is described by adding further HTML elements within the `body` element.

HTML tag:	`<img src="poleplot.png">`
element name:	`<img src="poleplot.png">` (`img`)
attribute:	`<img src="poleplot.png">` (`src="poleplot.png"`)
attribute name:	`<img src="poleplot.png">` (`src`)
attribute value:	`<img src="poleplot.png">` (`poleplot.png`)

There is a fixed set of valid HTML elements (Section 3.2.1 provides a list of some common elements) and each element has its own set of possible attributes.

Certain HTML elements are compulsory. An HTML document must include a `DOCTYPE` declaration and a single `html` element. Within the `html` element there must be a single `head` element and a single `body` element. Within the `head` element there must be a `title` element. Figure 3.1 shows a minimal piece of HTML code.

Section 3.2.1 describes each of the common elements in a little more detail, including any important attributes and which elements may be placed inside which other elements.

3.1.1 HTML comments

Comments in HTML code are anything within `<!--` and `-->`. All characters, including HTML tags, lose their special meaning within an HTML comment.

Table 3.1: Some common HTML entities.

Character	Description	Entity
<	less-than sign	<
>	greater-than sign	>
&	ampersand	&
π	Greek letter pi	π
μ	Greek letter mu	μ
€	Euro symbol	€
£	British pounds	£
©	copyright symbol	©

3.1.2 HTML entities

The less-than and greater-than characters used in HTML tags are special characters and must be escaped to obtain their literal meaning. These **escape sequences** in HTML are called **entities**. All entities start with an ampersand so the ampersand is also special and must be escaped. Entities provide a way to include some other special characters and symbols within HTML code as well. Table 3.1 shows some common HTML entities.

3.2 HTML semantics

The primary purpose of HTML tags is to specify the *structure* of a web page.

Elements are either **block-level** or **inline**. A block-level element is like a paragraph; it is a container that can be filled with other elements. Most block-level elements can contain any other sort of element. An inline element is like a word within a paragraph; it is a small component that is arranged with other components inside a container. An inline element usually only contains text.

The content of an element may be other elements or plain text. There is a limit on which elements may be nested within other elements (see Section 3.2.1).

3.2.1 Common HTML elements

This section briefly describes the important behavior, attributes, and rules for each of the common HTML elements.

<html>

The html element must contain exactly one head element followed by exactly one body element.

<head>

The head element is only allowed within the html element. It must contain exactly one title element. It may also contain link elements to refer to external CSS files and/or style elements for inline CSS rules. It has no attributes of interest.

<title>

The title element must be within the head element and must only contain text. This element provides information for the computer to use to identify the web page rather than for display, though it is often displayed in the title bar of the browser window. It has no attributes.

<link>

This is an empty element that must reside in the head element. It can be used to specify an external CSS file, in which case the important attributes are: rel, which should have the value "stylesheet"; href, which specifies the location of a file containing CSS code (can be a URL); and type, which should have the value "text/css". The media attribute may also be used to distinguish between a style sheet for display on "screen" as opposed to display in "print".

An example of a link element is shown below.

```
<link rel="stylesheet" href="csscode.css"
      type="text/css">
```

Other sorts of links are also possible, but are beyond the scope of this book.

<body>

The body element is only allowed within the html element. It should only contain one or more block-level elements, but most browsers will also allow inline elements. Various appearance-related attributes are possible, but CSS should be used instead.

<p>

This is a block-level element that can appear within most other block-level elements. It should only contain inline elements (words and images). The content is automatically typeset as a paragraph (i.e., the browser automatically decides where to break lines).

`<img>`

This is an empty, inline element (i.e., images are treated like words in a sentence). It can be used within almost any other element. Important attributes are `src`, to specify the file containing the image (this may be a URL, i.e., an image anywhere on the web), and `alt` to specify alternative text for non-graphical browsers.

`<a>`

The `a` element is known as an **anchor**. It is an inline element that can go inside any other element. It can contain any other inline element (except another anchor). Its important attributes are: `href`, which means that the anchor is a hypertext link and the value of the attribute specifies a destination (when the content of the anchor is clicked on, the browser navigates to this destination); and `name`, which means that the anchor is the destination for a hyperlink.

The value of an `href` attribute can be: a URL, which specifies a separate web page to navigate to; something of the form `#target`, which specifies an anchor within the same document that has an attribute `name="target"`; or a combination, which specifies an anchor within a separate document. For example, the following URL specifies the top of the W3C page for HTML 4.01.

```
href="http://www.w3.org/TR/html401/"
```

The URL below specifies the table of contents further down that web page.

```
href="http://www.w3.org/TR/html401/#minitoc"
```

`<h1>` ... `<h6>`

These are block-level elements that denote that the contents are a section heading. They can appear within almost any other block-level element, but can only contain inline elements. They have no attributes of interest.

These elements should be used to indicate the section structure of a document, not for their default display properties. CSS should be used to achieve the desired weight and size of the text in general.

`<table>`, `<tr>`, and `<td>`

A `table` element contains one or more `tr` elements, each of which contains one or more `td` elements (so `td` elements can only appear within tr elements, and tr elements can only appear within a table element). A table element may appear within almost any other block-level element. In particular, a table can be nested within the td element of another table.

The table element has a summary attribute to describe the table for non-graphical browsers. There are also attributes to control borders, background colors, and widths of columns, but CSS is the preferred way to control these features.

The tr element has attributes for the alignment of the contents of columns, including aligning numeric values on decimal points. The latter is important because it has no corresponding CSS property.

The td element also has alignment attributes for the contents of a column for one specific row, but these can be handled via CSS instead. However, there are several attributes specific to td elements, in particular, rowspan and colspan, which allow a single cell to spread across more than one row or column.

Unless explicit dimensions are given, the table rows and columns are automatically sized to fit their contents.

It is tempting to use tables to arrange content on a web page, but it is recommended to use CSS for this purpose instead. Unfortunately, the support for CSS in web browsers tends to be worse than the support for table elements, so it may not always be possible to use CSS for arranging content. This warning also applies to controlling borders and background colors via CSS.

The code below shows an example of a table with three rows and three columns and the image below the code shows what the result looks like in a web browser.

```
<table border="1">
    <tr>
        <td></td> <td>pacific</td> <td>eurasian</td>
    </tr>
    <tr>
        <td>min</td> <td>276</td> <td>258</td>
    </tr>
    <tr>
        <td>max</td> <td>283</td> <td>293</td>
    </tr>
</table>
```

	pacific	eurasian
min	276	258
max	283	293

It is also possible to construct more complex tables with separate `thead`, `tbody`, and `tfoot` elements to group rows within the table (i.e., these three elements can go inside a `table` element, with `tr` elements inside them).

`<hr>`

This is an empty element that produces a horizontal line. It can appear within almost any block-level element. It has no attributes of interest.

This entire element can be replaced by CSS control of borders.

`<br>`

This is an empty element that forces a new line or line-break. It can be put anywhere. It has no attributes of interest.

This element should be used sparingly. In general, text should be broken into lines by the browser to fit the available space.

`<ul>`, `<ol>`, and `<li>`

These elements create lists. The `li` element generates a list item and appears within either a `ul` element, for a bullet-point list, or an `ol` element, for a numbered list.

Anything can go inside an `li` element (i.e., you can make a list of text descriptions, a list of tables, or even a list of lists).

These elements have no attributes of interest. CSS can be used to control the style of the bullets or numbering and the spacing between items in the list.

The code below shows an ordered list with two items and the image below the code shows what the result looks like in a web browser.

```
<ol>
    <li>
    <p>
    Large bodies of water tend to
    absorb and release heat more slowly
    compared to large land masses.
    </p>
    </li>

    <li>
    <p>
    Temperatures vary differently over time
    depending on which hemisphere the pole
    is located in.
    </p>
    </li>
</ol>
```

File Edit View Go Bookmarks Tools Help

1. Large bodies of water tend to absorb and release heat more slowly compared to large land masses.
2. Temperatures vary differently over time depending on which hemisphere the pole is located in.

It is also possible to produce "definition" lists, where each item has a heading. Use a `dl` element for the overall list with a `dt` element to give the heading and a `dd` element to give the definition for each item.

`<pre>`
: This is a block-level element that displays any text content exactly as it appears in the source code. It is useful for displaying computer code or computer output. It has no attributes of interest.

  It is possible to have other elements within a `pre` element. Like the `hr` element, this element can usually be replaced by CSS styling.

`<div>` and `<span>`
: These are generic block-level and inline elements (respectively). They have no attributes of interest.

  These can be used as "blank" elements with no predefined appearance properties. Their appearance can then be fully specified via CSS. In theory, any other HTML element can be emulated using one of these elements and appropriate CSS properties. In practice, the standard HTML elements are more convenient for their default behavior and these elements are used for more exotic situations.

3.2.2 Common HTML attributes

Almost all elements may have a `class` attribute, so that a CSS style specified in the `head` element can be associated with that element. Similarly, all elements may have an `id` attribute, which can be used to associate a CSS style. The value of all `id` attributes within a piece of HTML code must be unique.

All elements may also have a `style` attribute, which allows "inline" CSS rules to be specified within the element's start tag.

3.3 Further reading

The W3C HTML 4.01 Specification
    `http://www.w3.org/TR/html401/`
    The formal and official definition of HTML. Quite technical.

Getting started with HTML
    by Dave Raggett
    `http://www.w3.org/MarkUp/Guide/`
    An introductory tutorial to HTML by one of the original designers of the language.

The Web Design Group's HTML 4 web site
    `http://htmlhelp.com/reference/html40/`
    A more friendly, user-oriented description of HTML.

The w3schools HTML Tutorial
    `http://www.w3schools.com/html/`
    A quick, basic tutorial-based introduction to HTML.

HTML Tidy
    `http://www.w3.org/People/Raggett/tidy/`
    A description of HTML Tidy, including links to online versions.

The W3C Validation Service
    `http://validator.w3.org/`
    A more sophisticated validation service than HTML Tidy.

4

CSS Reference

CSS

Cascading Style Sheets (CSS) is a language used to specify the appearance of web pages—fonts, colors, and how the material is arranged on the page.

CSS is run when it is linked to some HTML code (see Section 4.3) and that HTML code is run.

The information in this chapter describes CSS level 1, which is a W3C Recommendation.

4.1 CSS syntax

CSS code consists of one or more **rules**.

Each CSS rule consists of a **selector** and, within brackets, one or more **properties**.

The selector specifies which HTML elements the rule applies to and the properties control the way that those HTML elements are displayed. An example of a CSS rule is shown below:

```
a {
    color: white;
}
```

The code `a` is the selector and the property is `color`, with the value `white`.

selector:	a {
property name:	color: white;
property value:	color: white;
	}

Just as for HTML, it is important to check CSS code for correct syntax. Most browsers will silently ignore errors in CSS code. For example, if a CSS property is not spelled correctly, it will just appear not to be working in the browser. The W3C provides an online validation service (see Section 4.5).

4.2 CSS semantics

There are a number of ways to specify the CSS selector, which determines the HTML elements that will be affected by a specific rule.

4.2.1 CSS selectors

Within a CSS rule, the selector specifies which HTML elements will be affected by the rule.

Element selectors:

The selector is just the name of an HTML element. All elements of this type in the linked HTML code will be affected by the rule. An example is shown below:

```
a {
    color: white;
}
```

This rule will apply to *all* anchor (`a`) elements within the linked HTML code.

The same rule may be applied to more than one type of element at once, by specifying several element names, separated by commas. For example, the following rule would apply to *both* `a` elements *and* `p` elements.

```
p, a {
    color: white;
}
```

Contextual selectors:

The selector is several element names, separated by *spaces*. This allows a CSS rule to be applied to an element only when the element is contained within another type of element. For example, the following rule will only apply to anchors that are within paragraphs (*not*, for example, to anchors that are within headings).

```
p a {
    color: white;
}
```

Contrast this to the previous CSS specifier where the element names are separated by a comma.

Class selectors:

The selector contains a full stop (.) and the part after the full stop describes the name of a **class**. All elements that have a `class` attribute with the appropriate value will be affected by the rule. An example is shown below:

```
p.footer {
    font-style: italic;
}
```

This rule will apply to any paragraph (`p`) element that has the attribute `class="footer"`. It will *not* apply to other `p` elements. It will not apply to other HTML elements, even if they have the attribute `class="footer"`.

If no HTML element name is specified, the rule will apply to *all* HTML elements with the appropriate class. An example is shown below:

```
.figure {
    margin-left: auto;
    margin-right: auto;
}
```

This rule will apply to *any* HTML element that has the attribute `class="figure"`.

ID selectors:

The selector contains a hash character (#). The rule will apply to all elements that have an appropriate `id` attribute. This type of rule can be used to control the appearance of exactly one element. An example is shown below:

```
p#footer {
    font-style: italic;
}
```

This rule will apply to the paragraph (`p`) element that has the attribute `id="footer"`. There can only be one such element within a piece of HTML code because the `id` attribute must be unique for all elements. This means that the HTML element name is redundant and can be left out. The rule below has the same effect as the previous rule:

```
#footer {
    font-style: italic;
}
```

It is possible for CSS rules to conflict, i.e., for there to be more than one rule for the same element.

In general, a more specific rule, e.g., a class selector, will override a less specific one. Otherwise, the rule that is specified last wins. For example, if two CSS files are linked in the header of an HTML document and they both contain rules with the same selector, then the rule in the second file will override the rule in the first file.

Rules are also usually inherited, except where it would not make sense. For example, if there is a rule that makes the text italic for the `body` of an HTML document, then all `p` elements within the `body` will have italic text, unless another rule specifies otherwise. However, a rule controlling the margins of the `body` would not be applied to the margins of the `p` elements (i.e., the body would be indented within the page, but the paragraphs would not be indented within the body as well, unless they specified margins themselves).

4.2.2 CSS properties

This section describes some of the common CSS properties, including the values that each property can take.

`font-family:`

This property controls the overall font family (the general style) for text within an element. The value can be a generic font type, for example, `monospace` or `serif`, or it can be a specific font family name, for example, `Courier` or `Times`. If a specific font is specified, it is usually a good idea to also include (after a comma) a generic font as well in case the person viewing the result does not have the specific font on their computer. An example is shown below:

```
font-family:  Times, serif
```

This means that a Times font will be used if it is available; otherwise, the browser will choose a serif font that is available.

`font-style:`, `font-weight:`, and `font-size:`

These properties control the detailed appearance of text. The style can be `normal` or `italic`, the weight can be `normal` or `bold`, and the size can be `large` or `small`.

There are a number of relative values for size (they go down to `xx-small` and up to `xx-large`), but it is also possible to specify an absolute size, such as `24pt`.

`color:` and `background-color:`

These properties control the foreground color (e.g., for displaying text), and the background color for an element.

For specifying the color value, there are a few basic color names, e.g., `black`, `white`, `red`, `green`, and `blue`, but for anything else it is necessary to specify a red-green-blue (RGB) triplet. This consists of an amount of red, an amount of green, and an amount of blue. The amounts can be specified as percentages so that, for example, `rgb(0%, 0%, 0%)` is black and `rgb(100%, 100%, 100%)` is white, and Ferrari red is `rgb(83%, 13%, 20%)`.

`text-align:`

This property controls the alignment of text within an element, with possible values `left`, `right`, `center`, or `justify`. This property only makes sense for block-level elements.

`width:` and `height:`

These properties provide explicit control of the width or height of an element. By default, these are the amount of space required for the element. For example, a paragraph of text expands to fill the width of the page and uses as many lines as necessary, while an image has an intrinsic size (number of pixels in each direction).

Explicit widths or heights can be either percentages (of the parent element) or an absolute value. Absolute values must include a unit, e.g., `in` for inches, `cm` for centimeters, or `px` for pixels. For example, within a web page that is 800 pixels wide on a screen that has a resolution of 100 dots-per-inch (dpi), to make a paragraph of text half the width of the page, the following three specifications are identical:

```
p { width: 50% }

p { width: 4in }

p { width: 400px }
```

`border-width:`, `border-style:`, and `border-color:`

These properties control the appearance of borders around an element. Borders are only drawn if the `border-width` is greater than zero. Valid border styles include `solid`, `double`, and `inset` (which produces a fake 3D effect).

These properties affect all borders, but there are other properties that affect only the top, left, right, or bottom border of an element. For example, it is possible to produce a horizontal line at the top of a paragraph by using just the `border-top-width` property.

CSS

`margin:`
: This property controls the space around the outside of the element (between this element and neighboring elements). The size of margins can be expressed using units, as for the `width` and `height` properties.

  This property affects all margins (top, left, right, and bottom). There are properties, e.g., `margin-top`, for controlling individual margins instead.

`padding:`
: This property controls the space between the border of the element and the element's contents. Values are specified as they are for margins. There are also specific properties, e.g., `padding-top`, for individual control of the padding on each side of the element.

`display:`
: This property controls how the element is arranged relative to other elements. A value of `block` means that the element is like a self-contained paragraph (typically, with an empty line before it and an empty line after it). A value of `inline` means that the element is just placed beside whatever was the previous element (like words in a sentence). The value `none` means that the element is not displayed at all.

  Most HTML elements are either intrinsically block-level or inline, so some uses of this property will not make sense.

`whitespace:`
: This property controls how whitespace in the content of an element is treated. By default, any amount of whitespace in HTML code collapses down to just a single space when displayed, and the browser decides when a new line is required. A value of `pre` for this property forces all whitespace within the content of an element to be displayed (especially all spaces and all new lines).

`float:`
: This property can be used to allow text (or other inline elements) to wrap around another element (such as an image). The value `right` means that the element (e.g., image) "floats" to the right of the web page and other content (e.g., text) will fill in the gap to the left. The value `left` works analogously.

`clear:`
: This property controls whether floating elements are allowed beside an element. The value `both` means that the element will be placed below any previous floating elements. This can be used to have the effect of turning off text wrapping.

4.3 Linking CSS to HTML

CSS code can be linked to HTML code in one of three ways. These are described below in increasing order of preference.

Inline CSS:

The simplest approach is to include CSS code within the `style` attribute of an HTML element. An example is shown below:

```
<p style="font-style: italic">
```

Here, CSS is used to control the appearance of text within this paragraph only.

This approach is actively discouraged because it leads to many copies of the same CSS code within a single piece of HTML code.

Embedded CSS:

It is also possible to include CSS code within a `style` element within the `head` element of HTML code. An example of this is shown below:

```
<html>
    <head>
        <style>
            p {
                font-style: italic;
            }
        </style>
    ...
```

In this case, the appearance of text within *all* paragraphs is controlled with a single CSS rule.

This approach is better because rules are not attached to each individual HTML element. However, this is still not ideal because any reuse of the CSS code with other HTML code requires copying the CSS code (which violates the DRY principle).

External CSS:

The third approach is to write CSS code in a separate file and refer to the CSS file from the HTML code, using a `link` element within the `head` element. An example is shown below:

```
<link rel="stylesheet" href="csscode.css"
      type="text/css">
```

This line would go within a file of HTML code and it refers to CSS code within a file called `csscode.css`.

This approach allows, for example, control over the appearance of text within *all* paragraphs for *multiple* HTML files at once.

4.4 CSS tips

In some cases, it is not immediately obvious how to perform some basic formatting tasks with CSS. This section provides pointers for a few of the most common of these situations.

Indenting:

In HTML, whitespace and line breaks in the HTML *source* are generally ignored, so adding spaces or tabs at the start of the line in HTML code has no effect on the output.

Indenting text away from the left hand edge of the page can be achieved with CSS using either `margin` or `padding` properties.

Centering:

An element may be centered on the page (more precisely, within its containing element) by setting *both* the `margin-left` and `margin-right` properties to `auto`.

Floating text:

It is easy to start having text flow around another element, such as an image, by specifying something like `float: right` for the image element. However, it is not as obvious how to *stop* elements from floating next to each other.

This is what the `clear` property is useful for. In particular, `clear: both` is useful to turn off all floating for subsequent elements.

Another way to get nice results is to make use of other people's CSS code. For example, the Yahoo! User Interface Library (YUI)[1] provides a variety of page layouts via CSS, though there is still the learning curve of how to use YUI itself.

[1] `http://developer.yahoo.com/yui/`

4.5 Further reading

The W3C CSS Level 1 Specification

`http://www.w3.org/TR/CSS1`

The formal and official definition of CSS (level 1). Quite technical.

Adding a Touch of Style

by Dave Raggett

`http://www.w3.org/MarkUp/Guide/Style.html`

An introductory tutorial to CSS by one of the original designers of HTML.

The Web Design Group's CSS web site

`http://htmlhelp.com/reference/css/`

A more friendly, user-oriented description of CSS.

The w3schools CSS Tutorial

`http://www.w3schools.com/css/`

A tutorial-based introduction to CSS.

The W3C CSS Validation Service

`http://jigsaw.w3.org/css-validator/`

A syntax checker for CSS code.

CSS Zen Garden

`http://www.csszengarden.com/`

A web site that demonstrates and evangelizes the flexibility and power of CSS beautifully.

5

Data Storage

Chapter 2 dealt with the ideas and tools that are necessary for producing computer code. In Chapter 2, we were concerned with how we communicate our instructions to the computer.

In this chapter, we move over to the computer's view of the world for a while and investigate how the computer takes information and stores it.

There are several good reasons why researchers need to know about data storage options. One is that we may not have control over the format in which data is given to us. For example, data from NASA's Live Access Server is in a format decided by NASA and we are unlikely to be able to convince NASA to provide it in a different format. This says that we must know about different formats in order to gain access to data.

Another common situation is that we may have to transfer data between different applications or between different operating systems. This effectively involves temporary data storage, so it is useful to understand how to select an appropriate storage format.

It is also possible to be involved in deciding the format for archiving a data set. There is no overall best storage format; the correct choice will depend on the size and complexity of the data set and what the data set will be used for. It is necessary to gain an overview of the relative merits of all of the data storage options in order to be able to make an informed decision for any particular situation.

In this chapter, we will see a number of different data storage options and we will discuss the strengths and weaknesses of each.

How this chapter is organized

This chapter begins with a case study that is used to introduce some of the important ideas and issues that arise when data is stored electronically. The purpose of this section is to begin thinking about the problems that can arise and to begin establishing some criteria that we can use to compare different data storage options.

The remaining sections each describe one possible approach to data storage: Section 5.2 discusses plain text formats, Section 5.3 discusses binary formats,

Section 5.4 looks briefly at the special case of spreadsheets, Section 5.5 describes XML, and Section 5.6 addresses relational databases.

In each case, there is an explanation of the relevant technology, including a discussion of how to use the technology in an appropriate way, and there is a discussion of the strengths and weaknesses of each storage option.

5.1 Case study: YBC 7289

YBC 7289. Photo by Bill Casselman.

Some of the earliest known examples of recorded information come from Mesopotamia, which roughly corresponds to modern-day Iraq, and date from around the middle of the fourth millenium BC. The writing is called *cuneiform*, which refers to the fact that marks were made in wet clay with a wedge-shaped stylus.

A particularly famous mathematical example of cuneiform is the clay tablet known as YBC 7289.

This tablet is inscribed with a set of numbers using the Babylonian sexagesimal (base-60) system. In this system, an angled symbol, <, represents the value 10 and a vertical symbol, |, represents the value 1. For example, the value 30 is written (roughly) like this: <<<. This value can be seen along the top-left edge of YBC 7289 (see Figure 5.1).

The markings across the center of YBC 7289 consist of four digits: |, <<||||, <<<<<|, and <. Historians have suggested that these markings represent an estimate of the length of the diagonal of a unit square, which has a true value of $\sqrt{2} = 1.41421356$ (to eight decimal places). The decimal interpretation of the sexagesimal digits is $1 + \frac{24}{60} + \frac{51}{3600} + \frac{10}{216000} = 1.41421296$, which is amazingly close to the true value, considering that YBC 7289 has been dated to around 1600 BC.

Figure 5.1: Clay tablet YBC 7289 with an overlay to emphasize the markings on its surface. The cuneiform inscriptions demonstrate the derivation of the square root of 2.

Storing YBC 7289

What we are going to do with this ancient clay tablet is to treat it as information that needs to be stored electronically.

The choice of a clay tablet for recording the information on YBC 7289 was obviously a good one in terms of the durability of the storage medium. Very few electronic media today have an expected lifetime of several thousand years. However, electronic media do have many other advantages.

The most obvious advantage of an electronic medium is that it is very easy to make copies. The curators in charge of YBC 7289 would no doubt love to be able to make identical copies of such a precious artifact, but truly identical copies are only really possible for electronic information.

This leads us to the problem of how we produce an electronic record of the tablet YBC 7289. We will consider a number of possibilities in order to introduce some of the issues that will be important when discussing various data storage alternatives throughout this chapter.

A straightforward approach to storing the information on this tablet would be to write a simple textual description of the tablet.

```
YBC 7289 is a clay tablet with various cuneiform marks on it
that describe the relationship between the length of the
diagonal and the length of the sides of a square.
```

This approach has the advantages that it is easy to create and it is easy for a human to access the information. However, when we store electronic information, we should also be concerned about whether the information is easily accessible for computer software. This essentially means that we should supply clear labels so that individual pieces of information can be retrieved easily. For example, the label of the tablet is something that might be used to identify this tablet from all other cuneiform artifacts, so the label information should be clearly identified.

```
label: YBC 7289
description: A clay tablet with various cuneiform marks on it
that describe the relationship between the length of the
diagonal and the length of the sides of a square.
```

Thinking about what sorts of questions will be asked of the data is a good way to guide the design of data storage. Another sort of information that people might go looking for is the set of cuneiform markings that occur on the tablet.

The markings on the tablet are numbers, but they are also symbols, so it would probably be best to record both numeric and textual representations. There are three sets of markings and three values to record for each set; a common way to record this sort of information is with a row of information per set of markings, with three columns of values on each row.

```
   <<<                      30            30
     | <<|||| <<<<<| <       1 24 51 10    2.41421296
<<<<|| <<||||| <<<|||||     42 25 35      42.4263889
```

When storing the lines of symbols and numbers, we have spaced out the information so that it is easy, for a human, to see where one sort of value ends and another begins. Again, this information is even more important for the computer. Another option is to use a special character, such as a comma, to indicate the start/end of separate values.

```
values:
cuneiform,sexagesimal,decimal
<<<,30,30
| <<|||| <<<<<| <,1 24 51 10,2.41421296
<<<<|| <<||||| <<<|||||,42 25 35,42.4263889
```

Something else we should add is information about how the values relate to each other. Someone who is unfamiliar with Babylonian history may have difficulty realizing how the three values on each line actually correspond to each other. This sort of encoding information is essential **metadata**—information about the data values.

```
encoding: In cuneiform, a '<' stands for 10 and
a '|' stands for 1.  Sexagesimal values are base 60, with
a sexagesimal point after the first digit;  the first digit
represents ones, the second digit is sixtieths, the third
is three-thousand six-hundredths, and the fourth is two
hundred and sixteen thousandths.
```

The position of the markings on the tablet and the fact that there is also a square, with its diagonals inscribed, are all important information that contribute to a full understanding of the tablet. The best way to capture this information is with a photograph.

In many fields, data consist not just of numbers, but also pictures, sounds, and video. This sort of information creates additional files that are not easily incorporated together with textual or numerical data. The problem becomes not only how to store each individual representation of the information, but also how to *organize* the information in a sensible way. Something that we could do in this case is include a reference to a file containing a photograph of the tablet.

```
photo: ybc7289.png
```

Information about the source of the data may also be of interest. For example, the tablet has been dated to sometime between 1800 BC and 1600 BC. Little is known of its rediscovery, except that it was acquired in 1912 AD by an agent of J. P. Morgan, who subsequently bequeathed it to Yale University. This sort of **metadata** is easy to record as a textual description.

```
medium: clay tablet
history: Created between 1800 BC and 1600 BC, purchased by
J.P. Morgan 1912, bequeathed to Yale University.
```

The YBC in the tablet's label stands for the Yale Babylonian Collection. This tablet is just one item within one of the largest collections of cuneiforms in the world. In other words, there are a lot of other sources of data very similar to this one.

This has several implications for how we should store information about YBC 7298. First of all, we should store the information about this tablet in the same way that information is stored for other tablets in the collection so that, for example, a researcher can search for all tablets created in a certain time period. We should also think about the fact that some of the information that we have stored for YBC 7289 is very likely to be in common with all items in the collection. For example, the explanation of the sexagesimal system will be the same for other tablets from the same era. With this in mind, it does not make sense to record the encoding information for every single tablet. It would make sense to record the encoding information once, perhaps in a separate file, and just refer to the appropriate encoding information within the record for an individual tablet.

A complete version of the information that we have recorded so far might look like this:

```
label: YBC 7289
description: A clay tablet with various cuneiform marks on it
that describe the relationship between the length of the
diagonal and the length of the sides of a square.
photo: ybc7289.png
medium: clay tablet
history: Created between 1800 BC and 1600 BC, purchased by
J.P. Morgan 1912, bequeathed to Yale University.
encoding:  sexagesimal.txt
values:
cuneiform,sexagesimal,decimal
<<<,30,30
| <<|||| <<<<<| <,1 24 51 10,2.41421296
<<<<|| <<||||| <<<|||||,42 25 35,42.4263889
```

Is this the best possible way to store information about YBC 7289? Almost certainly not. Some problems with this approach include the fact that storing information as text is often not the most efficient approach and the fact that it would be difficult and slow for a computer to extract individual pieces of information from a free-form text format like this. However, the choice of an appropriate format also depends on how the data will be used.

The options discussed so far have only considered a couple of the possible *text* representations of the data. Another whole set of options to consider is *binary* formats. For example, the photograph and the text and numeric information could all be included in a single file. The most likely solution in practice is that this information resides in a *relational database* of information that describes the entire Yale Babylonian Collection.

This chapter will look at the decisions involved in choosing a format for storing information, we will discuss a number of standard data storage formats, and we will acquire the technical knowledge to be able to work with the different formats.

We start in Section 5.2 with plain text formats. This is followed by a discussion of binary formats in Section 5.3, and in Section 5.4, we look at the special case of spreadsheets. In Section 5.5, we look at XML, a computer language for storing data, and in Section 5.6, we discuss relational databases.

5.2 Plain text formats

The simplest way to store information in computer memory is as a single file with a plain text format.

Plain text files can be thought of as the lowest common denominator of storage formats; they might not be the most efficient or sophisticated solution, but we can be fairly certain that they will get the job done.

The basic conceptual structure of a plain text format is that the data are arranged in rows, with several values stored on each row.

It is common for there to be several rows of general information about the data set, or **metadata**, at the start of the file. This is often referred to as a file **header**.

A good example of a data set in a plain text format is the surface temperature data for the Pacific Pole of Inaccessibility (see Section 1.1). Figure 5.2 shows how we would normally see this sort of plain text file if we view it in a text editor or a web browser.

This file has 8 lines of metadata at the start, followed by 48 lines of the core data values, with 2 values, a date and a temperature, on each row.

There are two main sub-types of plain text format, which differ in how separate values are identified within a row:

Delimited formats:

In a delimited format, values within a row are separated by a special character, or **delimiter**. For example, it is possible to view the file

```
             VARIABLE : Mean TS from clear sky composite (kelvin)
             FILENAME : ISCCPMonthly_avg.nc
             FILEPATH : /usr/local/fer_data/data/
             SUBSET   : 48 points (TIME)
             LONGITUDE: 123.8W(-123.8)
             LATITUDE : 48.8S
                       123.8W
                        23
16-JAN-1994 00 /  1:  278.9
16-FEB-1994 00 /  2:  280.0
16-MAR-1994 00 /  3:  278.9
16-APR-1994 00 /  4:  278.9
16-MAY-1994 00 /  5:  277.8
16-JUN-1994 00 /  6:  276.1
...
```

Figure 5.2: The first few lines of the plain text output from the Live Access Server for the surface temperature at Point Nemo. This is a reproduction of Figure 1.2.

in Figure 5.2 as a delimited format, where each line after the header consists of two fields separated by a colon (the character ':' is the delimiter). Alternatively, if we used whitespace (one or more spaces or tabs) as the delimiter, there would be five fields, as shown below.

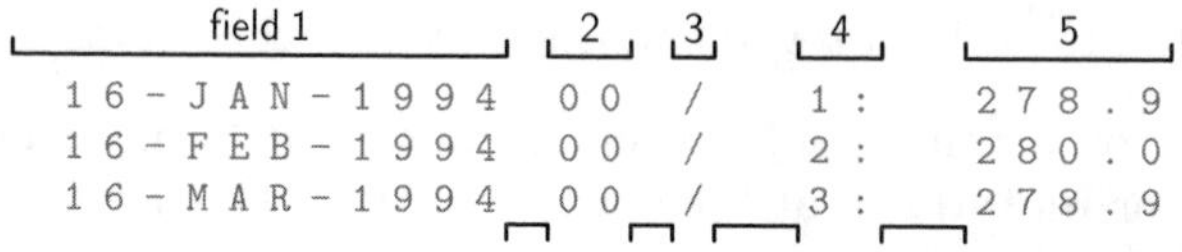

Fixed-width formats:

In a fixed-width format, each value is allocated a fixed number of characters within every row. For example, it is possible to view the file in Figure 5.2 as a fixed-width format, where the first value uses the first 20 characters and the second value uses the next 8 characters. Alternatively, there are five values on each row using 12, 3, 2, 6, and 5 characters respectively, as shown in the diagram below.

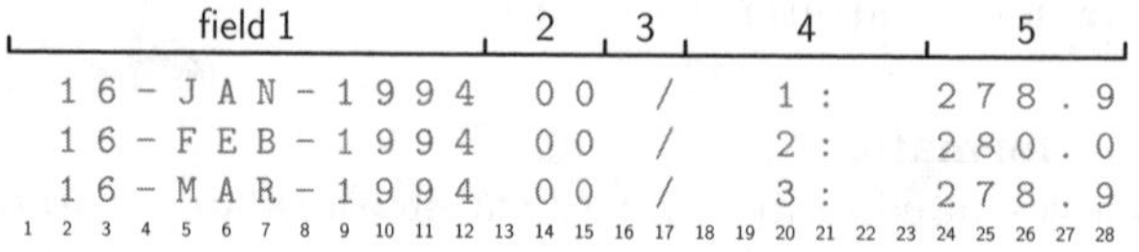

At the lowest level, the primary characteristic of a plain text format is that all of the information in the file, even numeric information, is stored as text.

We will spend the next few sections at this lower level of detail because it will be helpful in understanding the advantages and disadvantages of plain text formats for storing data, and because it will help us to differentiate plain text formats from *binary* formats later on in Section 5.3.

The first things we need to establish are some fundamental ideas about computer memory.

5.2.1 Computer memory

The most fundamental unit of computer memory is the **bit**. A bit can be a tiny magnetic region on a hard disk, a tiny dent in the reflective material on a CD or DVD, or a tiny transistor on a memory stick. Whatever the physical implementation, the important thing to know about a bit is that, like a switch, it can only take one of two values: it is either "on" or "off".

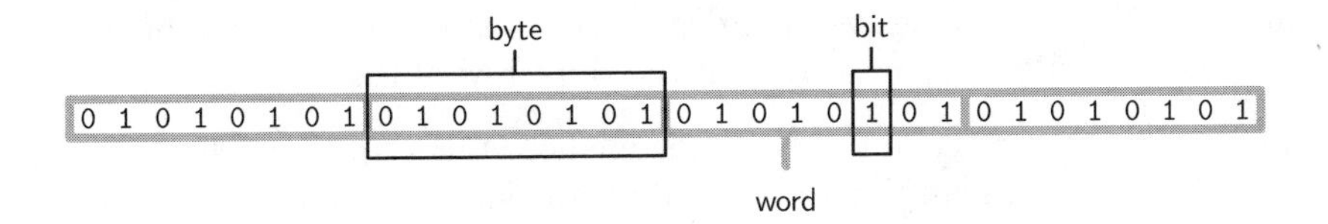

A collection of 8 bits is called a **byte** and (on the majority of computers today) a collection of 4 bytes, or 32 bits, is called a **word**.

5.2.2 Files and formats

A **file** is simply a block of computer memory.

A file can be as small as just a few bytes or it can be several gigabytes in size (thousands of millions of bytes).

A file **format** is a way of interpreting the bytes in a file. For example, in the simplest case, a **plain text** format means that each byte is used to represent a single character.

In order to visualize the idea of file formats, we will display a block of memory in the format shown below. This example shows the first 24 bytes from the PDF file for this book.

```
 0  :  00100101 01010000 01000100 01000110  |  %PDF
 4  :  00101101 00110001 00101110 00110100  |  -1.4
 8  :  00001010 00110101 00100000 00110000  |  .5 0
12  :  00100000 01101111 01100010 01101010  |   obj
16  :  00001010 00111100 00111100 00100000  |  .<<
20  :  00101111 01010011 00100000 00101111  |  /S /
```

This display has three columns. On the left is a byte offset that indicates the memory location within the file for each row. The middle column displays the raw memory contents of the file, which is just a series of 0's and 1's. The right hand column displays an interpretation of the bytes. This display is split across several rows just so that it will fit onto the printed page. A block of computer memory is best thought of as one long line of 0's and 1's.

In this example, we are interpreting each byte of memory as a single character, so for each byte in the middle column, there is a corresponding character in the right-hand column. As specific examples, the first byte, 00100101, is being interpreted as the percent character, %, and and the second byte, 01010000, is being interpreted as the letter P.

In some cases, the byte of memory does not correspond to a printable character, and in those cases we just display a full stop. An example of this is byte number nine (the first byte on the third row of the display).

Because the binary code for computer memory takes up so much space, we will also sometimes display the central raw memory column using hexadecimal (base 16) code rather than binary. In this case, each byte of memory is just a pair of hexadecimal digits. The first 24 bytes of the PDF file for this book are shown again below, using hexadecimal code for the raw memory.

```
 0  :  25 50 44 46 2d 31 2e 34 0a 35 20 30  |  %PDF-1.4.5 0
12  :  20 6f 62 6a 0a 3c 3c 20 2f 53 20 2f  |   obj.<< /S /
```

5.2.3 Case study: Point Nemo (continued)

We will now look at a low level at the surface temperature data for the Pacific Pole of Inaccessibility (see Section 1.1), which is in a plain text format. To emphasize the format of this information in computer memory, the first 48 bytes of the file are displayed below. This display should be compared with Figure 5.2, which shows what we would normally see when we view the plain text file in a text editor or web browser.

```
0   :  20 20 20 20 20 20 20 20 20 20 20 20  |
12  :  20 56 41 52 49 41 42 4c 45 20 3a 20  |   VARIABLE :
24  :  4d 65 61 6e 20 54 53 20 66 72 6f 6d  |  Mean TS from
36  :  20 63 6c 65 61 72 20 73 6b 79 20 63  |   clear sky c
```

This display clearly demonstrates that the Point Nemo information has been stored as a series of characters. The empty space at the start of the first line is a series of 13 spaces, with each space stored as a byte with the hexadecimal value `20`. The letter `V` at the start of the word `VARIABLE` has been stored as a byte with the value `56`.

To further emphasize the character-based nature of a plain text format, another part of the file is shown below as raw computer memory, this time focusing on the part of the file that contains the core data—the dates and temperature values.

```
336  :  20 31 36 2d 4a 41 4e 2d 31 39 39 34 20 30 30  |   16-JAN-1994 00
351  :  20 2f 20 20 31 3a 20 20 32 37 38 2e 39 0d 0a  |   /  1:  278.9..
366  :  20 31 36 2d 46 45 42 2d 31 39 39 34 20 30 30  |   16-FEB-1994 00
381  :  20 2f 20 20 32 3a 20 20 32 38 30 2e 30 0d 0a  |   /  2:  280.0..
```

The second line of this display shows that the *number* `278.9` is stored in this file as five characters—the digits 2, 7, 8, followed by a full stop, then the digit 9—with one byte per character. Another small detail that may not have been clear from previous views of these data is that each line starts with a space, represented by a byte with the value `20`.

We will contrast this sort of format with other ways of storing the information later in Section 5.3. For now, we just need to be aware of the simplicity of the memory usage in such a plain text format and the fact that *everything* is stored as a series of characters in a plain text format.

The next section will look at why these features can be both a blessing and a curse.

5.2.4 Advantages and disadvantages

The main advantage of plain text formats is their simplicity: we do not require complex software to create or view a text file and we do not need esoteric skills beyond being able to type on a keyboard, which means that it is easy for people to view and modify the data.

The simplicity of plain text formats means that virtually all software packages can read and write text files and plain text files are portable across different computer platforms.

The main *disadvantage* of plain text formats is also their simplicity. The basic conceptual structure of rows of values can be very inefficient and inappropriate for data sets with any sort of complex structure.

The low-level format of storing everything as characters, with one byte per character, can also be very inefficient in terms of the amount of computer memory required.

Consider a data set collected on two families, as depicted in Figure 5.3. What would this look like as a plain text file, with one row for all of the information about each person in the data set? One possible fixed-width format is shown below. In this format, each row records the information for one person. For each person, there is a column for the father's name (if known), a column for the mother's name (if known), the person's own name, his or her age, and his or her gender.

```
                 John       33  male
                 Julia      32  female
John      Julia  Jack        6  male
John      Julia  Jill        4  female
John      Julia  John jnr    2  male
                 David      45  male
                 Debbie     42  female
David     Debbie Donald     16  male
David     Debbie Dianne     12  female
```

This format for storing these data is not ideal for two reasons. Firstly, it is not efficient; the parent information is repeated over and over again. This repetition is also undesirable because it creates opportunities for errors and inconsistencies to creep in. Ideally, each individual piece of information would be stored exactly once; if more than one copy exists, then it is possible for the copies to disagree. The DRY principle (Section 2.7) applies to data as well as code.

The second problem is not as obvious, but is arguably much more important. The fundamental structure of most plain text file formats means that each line of the file contains exactly one record or case in the data set. This works well when a data set only contains information about one type of object, or, put another way, when the data set itself has a "flat" structure.

The data set of family members does not have a flat structure. There is information about two different types of object, parents and children, and these objects have a definite relationship between them. We can say that

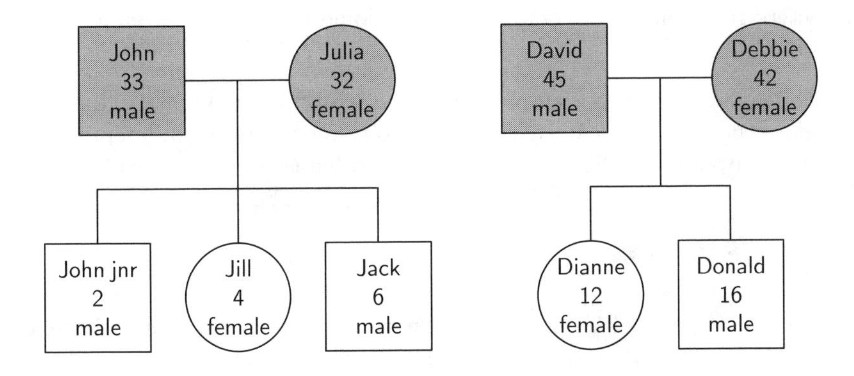

Figure 5.3: An example of hierarchical data: a family tree containing data on parents (grey) and children (white).

the data set is **hierarchical** or **multi-level** or **stratified** (as is obvious from the view of the data in Figure 5.3). Any data set that is obtained using a non-trivial study design is likely to have a hierarchical structure like this.

In other words, a plain text file format does not allow for sophisticated **data models**. A plain text format is unable to provide an appropriate representation of a complex data structure. Later sections will provide examples of storage formats that are capable of storing complex data structures.

Another major weakness of free-form text files is the lack of information *within the file itself* about the structure of the file. For example, plain text files do not usually contain information about which special character is being used to separate fields in a delimited file, or any information about the widths of fields with a fixed-width format. This means that the computer cannot automatically determine where different fields are within each row of a plain text file, or even how many fields there are.

A fixed-width format avoids this problem, but enforcing a fixed length for fields can create other difficulties if we do not know the maximum possible length for all variables. Also, if the values for a variable can have very different lengths, a fixed-width format can be inefficient because we store lots of empty space for short values.

The simplicity of plain text files make it easy for a computer to read a file as a series of characters, but the computer cannot easily distinguish individual data values from the series of characters. Even worse, the computer has no way of telling what sort of data is stored in each field. Does the series of characters represent a number, or text, or some more complex value such as a date?

In practice, a human must supply additional information about a plain text file before the computer can successfully determine where the different fields are within a plain text file *and* what sort of value is stored in each field.

5.2.5 CSV files

The Comma-Separated Value (CSV) format is a special case of a plain text format. Although not a formal standard, CSV files are very common and are a quite reliable plain text delimited format that at least solves the problem of where the fields are in each row of the file.

The main rules for the CSV format are:

Comma-delimited:
Each field is separated by a comma (i.e., the character `,` is the delimiter).

Double-quotes are special:
Fields containing commas must be surrounded by double-quotes (i.e., the `"` character is special).

Double-quote escape sequence:
Fields containing double-quotes must be surrounded by double-quotes *and* each embedded double-quote must be represented using two double-quotes (i.e., within double-quotes, `""` is an escape sequence for a literal double-quote).

Header information
There can be a single header line containing the names of the fields.

CSV files are a common way to transfer data from a spreadsheet to other software.

Figure 5.4 shows what the Point Nemo temperature data might look like in a CSV format. Notice that most of the metadata cannot be included in the file when using this format.

5.2.6 Line endings

A common feature of plain text files is that data values are usually arranged in rows, as we have seen in Figures 5.2 and 5.4.

We also know that plain text files are, at the low level of computer memory,

```
date,temp
 16-JAN-1994,278.9
 16-FEB-1994,280
 16-MAR-1994,278.9
 16-APR-1994,278.9
 16-MAY-1994,277.8
 16-JUN-1994,276.1
 ...
```

Figure 5.4: The first few lines of the plain text output from the Live Access Server for the surface temperature at Point Nemo in Comma-Separated Value (CSV) format. On each line, there are two data values, a date and a temperature value, separated from each other by a comma. The first line provides a name for each column of data values.

just a series of characters.

How does the computer know where one line ends and the next line starts?

The answer is that the end of a line in a text file is indicated by a special character (or two special characters). Most software that we use to view text files does not explicitly show us these characters. Instead, the software just starts a new line.

To demonstrate this idea, two lines from the file `pointnemotemp.txt` are reproduced below (as they appear when viewed in a text editor or web browser).

```
16-JAN-1994 00 /   1:   278.9
16-FEB-1994 00 /   2:   280.0
```

The section of computer memory used to store these two lines is shown below.

```
336  :  20 31 36 2d 4a 41 4e 2d 31 39 39 34 20 30 30  |   16-JAN-1994 00
351  :  20 2f 20 20 31 3a 20 20 32 37 38 2e 39 0d 0a  |   /  1:  278.9..
366  :  20 31 36 2d 46 45 42 2d 31 39 39 34 20 30 30  |   16-FEB-1994 00
381  :  20 2f 20 20 32 3a 20 20 32 38 30 2e 30 0d 0a  |   /  2:  280.0..
```

The feature to look for is the section of two bytes immediately after each temperature value. These two bytes have the values `0d` and `0a`, and this is the special byte sequence that is used to indicate the end of a line in a plain text file.

As mentioned above, these bytes are not explicitly shown by most software that we use to view the text file. The software detects this byte sequence and starts a new line in response.

So why do we need to know about the special byte sequence at all? Because, unfortunately, this is not the only byte sequence used to signal the end of the line. The sequence `0d 0a` is common for plain text files that have been created on a Windows system, but for plain text files created on a Mac OS X or Linux system, the sequence is likely to be just `0a`.

Many computer programs will allow for this possibility and cope automatically, but it can be a source of problems. For example, if a plain text file that was created on Linux is opened using Microsoft Notepad, the entire file is treated as if it is one long row, because Notepad expects to see `0d 0a` for a line ending and the file will only contain `0a` at the end of each line.

5.2.7 Text encodings

We have identified two features of plain text formats so far: all data is stored as a series of characters *and* each character is stored in computer memory using a single byte.

The second part, concerning how a single character is stored in computer memory, is called a character **encoding**.

Up to this point we have only considered the simplest possible character encoding. When we only have the letters, digits, special symbols, and punctuation marks that appear on a standard (US) English keyboard, then we can use a single byte to store each character. This is called an ASCII encoding (American Standard Code for Information Interchange).

An encoding that uses a single byte (8 bits) per character can cope with up to 256 (2^8) different characters, which is plenty for a standard English keyboard.

Many other languages have some characters in common with English but also have accented characters, such as é and ö. In each of these cases, it is still possible to use an encoding that represents each possible character in the language with a single byte. However, the problem is that different encodings may use the same byte value for a different character. For example, in the Latin1 encoding, for Western European languages, the byte value `f1` represents the character ñ, but in the Latin2 encoding, for Eastern European languages, the byte value `f1` represents the character ń. Because of this ambiguity, it is important to know what encoding was used when the text was stored.

The situation is much more complex for written languages in some Asian and Middle Eastern countries that use several thousand different characters (e.g., Japanese Kanji ideographs). In order to store text in these languages, it is necessary to use a **multi-byte** encoding scheme where more than one byte is used to store each character.

UNICODE is an attempt to allow computers to work with all of the characters in all of the languages of the world. Every character has its own number, called a "code point", often written in the form `U+xxxxxx`, where every `x` is a hexadecimal digit. For example, the letter 'A' is `U+000041` and the letter 'ö' is `U+0000F6`.

There are two main "encodings" that are used to store a UNICODE code point in memory. UTF-16 always uses two bytes per character of text and UTF-8 uses one or more bytes, depending on which characters are stored. If the text is only ASCII, UTF-8 will only use one byte per character.

For example, the text "just testing" is shown below saved via Microsoft's Notepad with a plain text format, but using three different encodings: ASCII, UTF-16, and UTF-8.

```
0  :  6a 75 73 74 20 74 65 73 74 69 6e 67  |  just testing
```

The ASCII format contains exactly one byte per character.

```
 0  :  ff fe 6a 00 75 00 73 00 74 00 20 00  |  ..j.u.s.t. .
12  :  74 00 65 00 73 00 74 00 69 00 6e 00  |  t.e.s.t.i.n.
24  :  67 00                                |  g.
```

The UTF-16 format differs from the ASCII format in two ways. For every byte in the ASCII file, there are now two bytes, one containing the hexadecimal code we saw before followed by a byte containing all zeroes. There are also two additional bytes at the start. These are called a byte order mark (**BOM**) and indicate the order of the two bytes that make up each letter in the text, which is used by software when reading the file.

```
 0  :  ef bb bf 6a 75 73 74 20 74 65 73 74  |  ...just test
12  :  69 6e 67                             |  ing
```

The UTF-8 format is mostly the same as the ASCII format; each letter has only one byte, with the same binary code as before because these are all common English letters. The difference is that there are three bytes at the start to act as a BOM. Notepad writes a BOM like this at the start of UTF-8 files, but not all software does this.

In summary, a plain text format always stores all data values as a series of characters. However, the number of bytes used to store each character in computer memory depends on the character encoding that is used.

This encoding is another example of additional information that may have to be provided by a human before the computer can read data correctly from a plain text file, although many software packages will cope with different encodings automatically.

5.2.8 Case study: The Data Expo

The TIROS Operational Vertical Sounder (TOVS) instruments have been used to collect atmospheric data aboard National Oceanic and Atmospheric Administration (NOAA) satellites since 1978.

The American Statistical Association (ASA) holds an annual conference called the Joint Statistical Meetings (JSM).

One of the events sometimes held at this conference is a Data Exposition, where contestants are provided with a data set and must produce a poster demonstrating a comprehensive analysis of the data. For the Data Expo at the 2006 JSM,[1] the data were geographic and atmospheric measures that were obtained from NASA's Live Access Server (see Section 1.1).

The variables in the data set are: elevation, temperature (surface and air), ozone, air pressure, and cloud cover (low, mid, and high). With the exception of elevation, all variables are monthly averages, with observations for January 1995 to December 2000. The data are measured at evenly spaced geographic locations on a very coarse 24 by 24 grid covering Central America (see Figure 5.5).

The data were downloaded from the Live Access Server in a plain text format with one file for each variable, for each month; this produced 72 files per atmospheric variable, plus 1 file for elevation, for a total of 505 files. Figure 5.6 shows the start of one of the surface temperature files.

This data set demonstrates a number of advantages and limitations of a plain text format for storing data. First of all, the data is very straightforward to access because it does not need sophisticated software. It is also easy for

[1] `http://stat-computing.org/dataexpo/2006/`

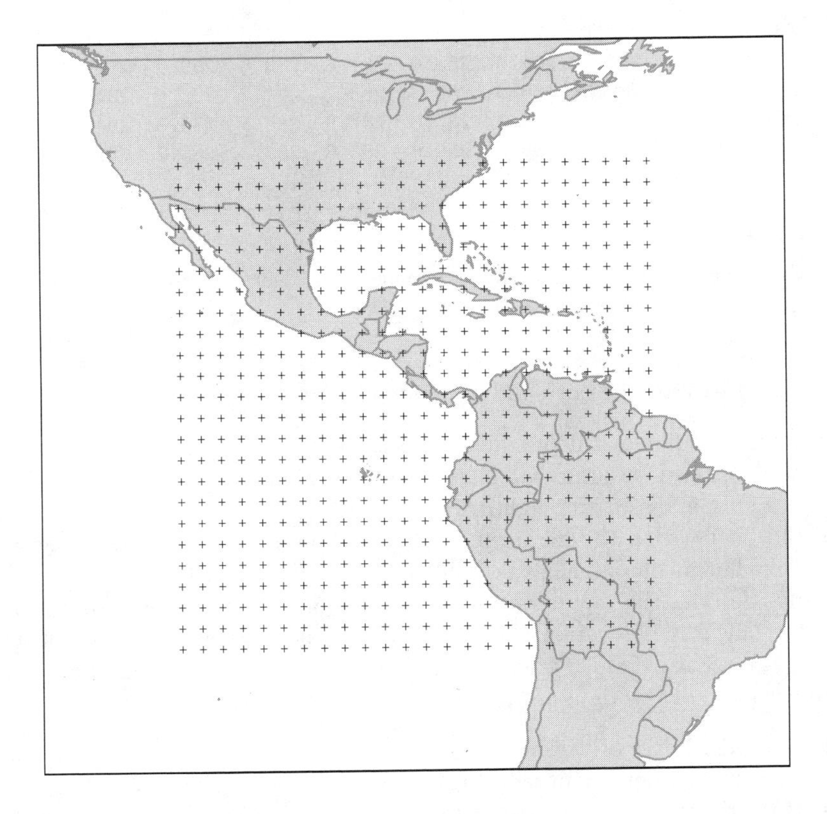

Figure 5.5: The geographic locations at which Live Access Server atmospheric data were obtained for the 2006 JSM Data Expo.

```
             VARIABLE : Mean TS from clear sky composite (kelvin)
             FILENAME : ISCCPMonthly_avg.nc
             FILEPATH : /usr/local/fer_dsets/data/
             SUBSET   : 24 by 24 points (LONGITUDE-LATITUDE)
             TIME     : 16-JAN-1995 00:00
              113.8W 111.2W 108.8W 106.2W 103.8W 101.2W 98.8W  ...
               27     28     29     30     31     32     33    ...
 36.2N / 51:  272.7  270.9  270.9  269.7  273.2  275.6  277.3  ...
 33.8N / 50:  279.5  279.5  275.0  275.6  277.3  279.5  281.6  ...
 31.2N / 49:  284.7  284.7  281.6  281.6  280.5  282.2  284.7  ...
 28.8N / 48:  289.3  286.8  286.8  283.7  284.2  286.8  287.8  ...
 26.2N / 47:  292.2  293.2  287.8  287.8  285.8  288.8  291.7  ...
 23.8N / 46:  294.1  295.0  296.5  286.8  286.8  285.2  289.8  ...
 ...
```

Figure 5.6: The first few lines of output from the Live Access Server for the surface temperature of the earth for January 1995, over a coarse 24 by 24 grid of locations covering Central America.

a human to view the data and understand what is in each file.

However, the file format provides a classic demonstration of the typical lack of standardized structure in plain text files. For example, the raw data values only start on the eighth line of the file, but there is no indication of that fact within the file itself. This is not an issue for a human viewing the file, but a computer has no chance of detecting this structure automatically. Second, the raw data are arranged in a matrix, corresponding to a geographic grid of locations, but again there is no inherent indication of this structure. For example, only a human can tell that the first 11 characters on each line of raw data are row labels describing latitude.

The "header" part of the file (the first seven lines) contains metadata, including information about which variable is recorded in the file and the units used for those measurements. This is very important and useful information, but again it is not obvious (for a computer) which bits are labels and which bits are information, let alone what sort of information is in each bit.

Finally, there is the fact that the data reside in 505 separate files. This is essentially an admission that plain text files are not suited to data sets with anything beyond a simple two-dimensional matrix-like structure. In this case, the temporal dimension—the fact that data are recorded at multiple time points—and the multivariate nature of the data—the fact that multiple variables are recorded—leads to there being separate files for each variable and for each time point. Having the data spread across many files creates issues in terms of the naming of files, for example, to ensure that all files from the same date, but containing different variables, can be easily located. There is also a reasonable amount of redundancy, with metadata and labels repeated many times over in different files.

We will look at another way to store this information in Section 5.6.5.

Recap

A plain text format is a simple, lowest-common-denominator storage format.

Data in a plain text format are usually arranged in rows, with several values on each row.

Values within a row are separated from each other by a delimiter or each value is allocated a fixed number of characters within a row.

The CSV format is a comma-delimited format.

All data values in a plain text file are stored as a series of characters. Even numbers are stored as characters.

Each character is stored in computer memory as one or two bytes.

The main problem with plain text files is that the file itself contains no information about where the data values are within the file and no information about whether the data values represent numbers or text, or something more complex.

5.3 Binary formats

In this section we will consider the option of storing data using a binary format. The purpose of this section is to understand the structure of binary formats and to consider the benefits and drawbacks of using a binary format to store data.

A binary format is a more complex storage solution than a plain text format, but it will typically provide faster and more flexible access to the data and use up less memory.

A file with a binary format is simply a block of computer memory, just like a file with a plain text format. The difference lies in how the bytes of computer memory are used.

In order to understand the difference, we need to learn a bit more about how computer memory works.

5.3.1 More on computer memory

We already know that one way to use a byte of computer memory is to interpret it as a character. This is what plain text formats do.

However, there are many other ways that we can make use of a byte of memory. One particularly straightforward use is to interpret the byte as an integer value.

A sequence of memory bits can be interpreted as a binary (base 2) number. For example, the byte `00000001` can represent the value 1, `00000010` can represent the value 2, `00000011` the value 3, and so on.

With 8 bits in a byte, we can store 256 (2^8) different integers (e.g., from 0 up to 255).

This is one way that binary formats can differ from plain text formats. If we want to store the value 48 in a plain text format, we must use two bytes, one for the digit `4` and one for the digit `8`. In a binary format, we could instead just use one byte and store the binary representation of 48, which is `00110000`.

One limitation of representing a number in computer memory this way is that we cannot store large integers, so in practice, a binary format will store an integer value using two bytes (for a maximum value of $2^{16} = 65,535$) or four bytes (maximum of $2^{32} = 4,294,967,295$). If we need to store negative numbers, we can use 1 bit for the sign. For example, using two bytes per integer, we can get a range of -32,767 to 32,767 ($\pm 2^{15} - 1$).

Another way to interpret a byte of computer memory is as a *real* number. This is not as straightforward as for integers, but we do retain the basic idea that each pattern of bits corresponds to a different number.

In practice, a minimum of 4 bytes is used to store real numbers. The following example of raw computer memory shows the bit patterns corresponding to the numbers 0.1 to 0.5, where each number is stored as a real number using 4 bytes per number.

```
 0  :  11001101 11001100 11001100 00111101  |  0.1
 4  :  11001101 11001100 01001100 00111110  |  0.2
 8  :  10011010 10011001 10011001 00111110  |  0.3
12  :  11001101 11001100 11001100 00111110  |  0.4
16  :  00000000 00000000 00000000 00111111  |  0.5
```

The correspondence between the bit patterns and the real numbers being stored is far from intuitive, but this is another common way to make use of bytes of memory.

As with integers, there is a maximum real number that can be stored using four bytes of memory. In practice, it is common to use eight bytes per real number, or even more. With 8 bytes per real number, the range is a little over $\pm 10^{307}$.

However, there is another problem with storing real numbers this way. Recall that, for integers, the first state can represent 0, the second state can represent 1, the third state can represent 2, and so on. With k bits, we can only go as high as the integer $2^k - 1$, but at least we know that we can account for all of the integers up to that point.

Unfortunately, we cannot do the same thing for real numbers. We could say that the first state represents 0, but what does the second state represent? 0.1? 0.01? 0.00000001? Suppose we chose 0.01, so the first state represents 0, the second state represents 0.01, the third state represents 0.02, and so on. We can now only go as high as $0.01 \times (2^k - 1)$, *and* we have missed all of the numbers between 0.01 and 0.02 (and all of the numbers between 0.02 and 0.03, and infinitely many others).

This is another important limitation of storing information on a computer: there is a limit to the **precision** that we can achieve when we store real numbers this way. Most real values cannot be stored exactly on a computer. Examples of this problem include not only exotic values such as transcendental numbers (e.g., π and e), but also very simple everyday values such as $\frac{1}{3}$ or even 0.1. Fortunately, this is not as dreadful as it sounds, because even if the exact value cannot be stored, a value very very close to the true value can be stored. For example, if we use eight bytes to store a real number, then we can store the distance of the earth from the sun to the nearest millimeter.

In summary, when information is stored in a binary format, the bytes of computer memory can be used in a variety of ways. To drive the point home even further, the following displays show *exactly the same* block of computer memory interpreted in three different ways.

First, we treat each byte as a single character.

```
0  :  74 65 73 74  |  test
```

Next, we interpret the memory as two, two-byte integers.

```
0  :  74 65 73 74  |  25972 29811
```

Finally, we can also interpret the memory as a single, four-byte real number.

```
0  :  74 65 73 74  |  7.713537e+31
```

Which one of these is the correct interpretation? It depends on which particular binary format has been used to store the data.

The characteristic feature of a binary format is that there is *not* a simple rule for determining how many bits or how many bytes constitute a basic unit of information.

It is necessary for there to be a description of the rules for the binary format that states what information is stored and how many bits or bytes are used for each piece of information.

Binary formats are consequently much harder to write software for, which results in there being less software available to do the job.

Given that a description is necessary to have any chance of reading a binary format, proprietary formats, where the file format description is kept private, are extremely difficult to deal with. Open standards become more important than ever.

The main point is that we require specific software to be able to work with data in a binary format. However, on the positive side, binary formats are generally more efficient and faster than text-based formats. In the next section, we will look at an example of a binary format and use it to demonstrate some of these ideas.

5.3.2 Case study: Point Nemo (continued)

In this example, we will use the Point Nemo temperature data (see Section 1.1) again, but this time with the data stored in a binary format called netCDF.

The first 60 bytes of the netCDF format file are shown in Figure 5.7, with each byte interpreted as a single character.

We will learn more about this format below. For now, the only point to make is that, while we can see that some of the bytes in the file appear to be text values, it is not clear from this raw display of the data, where the text data starts and ends, what values the non-text bytes represent, and what each value is for.

Compared to a plain text file, this is a complete mess and we need software that understands the netCDF format in order to extract useful values from the file.

```
 0  :  43 44 46 01 00 00 00 00 00 00 00 0a  |  CDF.........
12  :  00 00 00 01 00 00 00 04 54 69 6d 65  |  ........Time
24  :  00 00 00 30 00 00 00 00 00 00 00 00  |  ...0........
36  :  00 00 00 0b 00 00 00 02 00 00 00 04  |  ............
48  :  54 69 6d 65 00 00 00 01 00 00 00 00  |  Time........
```

Figure 5.7: The first 60 bytes of the netCDF format file that contains the surface temperatures at Point Nemo. The data are shown here as unstructured bytes to demonstrate that, without knowledge of the structure of the binary format, there is no way to determine where the different data values reside or what format the values have been stored in (apart from the fact that there is obviously some text values in the file).

5.3.3 NetCDF

The Point Nemo temperature data set has been stored in a format called network Common Data Form (netCDF).[2] In this section, we will learn a little bit about the netCDF format in order to demonstrate some general features of binary formats.

NetCDF is just one of a huge number of possible binary formats. It is a useful one to know because it is widely used and it is an open standard.

The precise bit patterns in the raw computer memory examples shown in this section are particular to this data set *and* the netCDF data format. Any other data set and any other binary format would produce completely different representations in memory. The point of looking at this example is that it provides a concrete demonstration of some useful general features of binary formats.

The first thing to note is that the structure of binary formats tends to be much more flexible than a text-based format. A binary format can use any number of bytes, in any order.

Like most binary formats, the structure of a netCDF file consists of **header** information, followed by the raw data itself. The header information includes information about how many data values have been stored, what sorts of values they are, and where within the file the header ends and the data values begin.

Figure 5.8 shows a structured view of the start of the netCDF format for

[2]`http://www.unidata.ucar.edu/software/netcdf/`

```
 0  :  43 44 46                      |  CDF

 3  :  01                            |  1

 4  :  00 00 00 00 00 00 00 0a       |   0 10
12  :  00 00 00 01 00 00 00 04       |   1  4

20  :  54 69 6d 65                   |  Time
```

Figure 5.8: The start of the header information in the netCDF file that contains the surface temperatures at Point Nemo, with the structure of the netCDF format revealed so that separate data fields can be observed. The first three bytes are characters, the fourth byte is a single-byte integer, the next sixteen bytes are four-byte integers, and the last four bytes are characters. This structured display should be compared to the unstructured bytes shown in Figure 5.7.

the Point Nemo temperature data. This should be contrasted with the unstructured view in Figure 5.7.

The netCDF file begins with three bytes that are interpreted as characters, specifically the three characters `C`, `D`, and `F`. This start to the file indicates that this is a netCDF file. The fourth byte in the file is a single-byte integer and specifies which version of netCDF is being used, in this case, version 1, or the "classic" netCDF format. This part of the file will be the same for any (classic) netCDF file.

The next sixteen bytes in the file are all four-byte integers and the four bytes after that are each single-byte characters again. This demonstrates the idea of binary formats, where values are packed into memory next to each other, with different values using different numbers of bytes.

Another classic feature of binary formats is that the header information contains pointers to the location of the raw data within the file and information about how the raw data values are stored. This information is not shown in Figure 5.8, but in this case, the raw data is located at byte 624 within the file and each temperature value is stored as an eight-byte real number.

Figure 5.9 shows the raw computer memory starting at byte 624 that contains the temperature values within the netCDF file. These are the first eight temperature values from the Point Nemo data set in a binary format.

In order to emphasize the difference in formats, the display below shows the raw computer memory from the *plain text* format of the Point Nemo

```
624  :  40 71 6e 66 66 66 66 66  |  278.9
632  :  40 71 80 00 00 00 00 00  |  280.0
640  :  40 71 6e 66 66 66 66 66  |  278.9
648  :  40 71 6e 66 66 66 66 66  |  278.9
656  :  40 71 5c cc cc cc cc cd  |  277.8
664  :  40 71 41 99 99 99 99 9a  |  276.1
672  :  40 71 41 99 99 99 99 9a  |  276.1
680  :  40 71 39 99 99 99 99 9a  |  275.6
...
```

Figure 5.9: The block of bytes within the Point Nemo netCDF file that contains the surface temperature values. Each temperature values is an eight-byte real number.

data set. Compare these five bytes, which store the number `278.9` as five characters, with the first eight bytes in Figure 5.9, which store the same number as an eight-byte real number.

```
359  :  32 37 38 2e 39  |  278.9
```

The section of the file shown in Figure 5.9 also allows us to discuss the issue of speed of access to data stored in a binary format.

The fundamental issue is that it is possible to *calculate* the location of a data value within the file. For example, if we want to access the fifth temperature value, 277.8, within this file, we know with certainty, because the header information has told us, that this value is 8 bytes long and it starts at byte number 656: the offset of 624, plus 4 times 8 bytes (the size of each temperature value).

Contrast this simple calculation with finding the fifth temperature value in a text-based format like the CSV format for the Point Nemo temperature data. The raw bytes representing the first few lines of the CSV format are shown in Figure 5.10.

The fifth temperature value in this file starts at byte 98 within the file, but there is no simple way to calculate that fact. The only way to find that value is to start at the beginning of the file and read one character at a time until we have counted six commas (the field separators in a CSV file). Similarly, because not all data values have the same length, in terms of number of bytes of memory, the end of the data value can only be found by continuing to read characters until we find the end of the line (in this file, the byte `0a`).

```
  0  :  64 61 74 65 2c 74 65 6d 70 0a 20  |  date,temp.
 11  :  31 36 2d 4a 41 4e 2d 31 39 39 34  |  16-JAN-1994
 22  :  2c 32 37 38 2e 39 0a 20 31 36 2d  |  ,278.9. 16-
 33  :  46 45 42 2d 31 39 39 34 2c 32 38  |  FEB-1994,28
 44  :  30 0a 20 31 36 2d 4d 41 52 2d 31  |  0. 16-MAR-1
 55  :  39 39 34 2c 32 37 38 2e 39 0a 20  |  994,278.9.
 66  :  31 36 2d 41 50 52 2d 31 39 39 34  |  16-APR-1994
 77  :  2c 32 37 38 2e 39 0a 20 31 36 2d  |  ,278.9. 16-
 88  :  4d 41 59 2d 31 39 39 34 2c 32 37  |  MAY-1994,27
 99  :  37 2e 38 0a 20 31 36 2d 4a 55 4e  |  7.8. 16-JUN
110  :  2d 31 39 39 34 2c 32 37 36 2e 31  |  -1994,276.1
```

Figure 5.10: A block of bytes from the start of the Point Nemo CSV file that contains the surface temperature values. Each character in the file occupies one byte.

The difference is similar to finding a particular scene in a movie on a DVD disc compared to a VHS tape.

In general, it is possible to jump directly to a specific location within a binary format file, whereas it is necessary to read a text-based format from the beginning and one character at a time. This feature of accessing binary formats is called **random access** and it is generally faster than the typical **sequential access** of text files.

This example is just provided to give a demonstration of how it is *possible* for access to be faster to data stored in a binary file. This does not mean that access speeds *are* always faster and it certainly does not mean that access speed should necessarily be the deciding factor when choosing a data format. In some situations, a binary format will be a good choice because data can be accessed quickly.

5.3.4 PDF documents

It is worthwhile briefly mentioning Adobe's Portable Document Format (PDF) because so many documents, including research reports, are now published in this binary format.

While PDF is not used directly as a data storage format, it is common for a report in PDF format to contain tables of data, so this is one way in which data may be received.

Unfortunately, extracting data from a table in a PDF document is not straightforward. A PDF document is primarily a description of how to *display* information. Any data values within a PDF document will be hopelessly entwined with information about how the data values should be displayed.

One simple way to extract the data values from a PDF document is to use the text selection tool in a PDF viewer, such as Adobe Reader, to cut-and-paste from the PDF document to a text format. However, there is no guarantee that data values extracted this way will be arranged in tidy rows and columns.

5.3.5 Other types of data

So far, we have seen how numbers and text values can be stored in computer memory. However, not all data values are simple numbers or text values.

For example, consider the date value March 1st 2006. How should that be stored in memory?

The short answer is that any value can be converted to either a number or a piece of text and we already know how to store those sorts of values. However, the decision of whether to use a numeric or a textual representation for a data value is not always straightforward.

Consider the problem of storing information on gender. There are (usually) only two possible values: `male` and `female`.

One way to store this information would be as text: "male" and "female". However, that approach would take up at least 4 to 6 bytes per observation. We could do better, in terms of the amount of memory used, by storing the information as an integer, with 1 representing male and 2 representing female, thereby only using as little as one byte per observation. We could do even better by using just a single bit per observation, with "on" representing male and "off" representing female.

On the other hand, storing the text value "male" is much less likely to lead to confusion than storing the number 1 or by setting a bit to "on"; it is much easier to remember or intuit that the text "male" corresponds to the male gender.

An ideal solution would be to store just the numbers, so that we use up less memory, *but* also record the mapping that relates the value 1 to the male gender.

Dates

The most common ways to store date values are as either text, such as "March 1 2006", or as a number, for example, the number of days since the first day of January, 1970.

The advantage of storing a date as a number is that certain operations, such as calculating the number of days between two dates, becomes trivial. The problem is that the stored value only makes sense if we also know the reference date.

Storing dates as text avoids the problem of having to know a reference date, but a number of other complications arise.

One problem with storing dates as text is that the format can differ between different countries. For example, the second month of the year is called February in English-speaking countries, but something else in other countries. A more subtle and dangerous problem arises when dates are written in formats like this: 01/03/06. In some countries, that is the first of March 2006, but in other countries it is the third of January 2006.

The solution to these problems is to use the international standard for expressing dates, ISO 8601. This standard specifies that a date should consist of four digits indicating the year, followed by two digits indicating the month, followed by two digits indicating the day, with dashes in between each component. For example, the first day of March 2006 is written: 2006-03-01.

Dates (a particular day) are usually distinguished from **date-times**, which specify not only a particular day, but also the hour, second, and even fractions of a second within that day. Date-times are more complicated to work with because they depend on location; mid-day on the first of March 2006 happens at different times for different countries (in different time zones). Daylight savings time just makes things worse.

ISO 8601 includes specifications for adding time information, including a time zone, to a date. As with simple dates, it is also common to store date-times as numbers, for example, as the number of *seconds* since the beginning of 1970.

Money

There are two major issues with storing monetary values. The first is that the currency should be recorded; NZ$1.00 is very different from US$1.00. This issue applies, of course, to any value with a unit, such as temperature, weight, and distances.

The second issue with storing monetary values is that values need to be recorded exactly. Typically, we want to keep values to exactly two decimal places at all times. Monetary data may be stored in a special format to allow for this emphasis on accuracy.

Metadata

In most situations, a single data value in isolation has no inherent meaning. We have just seen several explicit examples: a gender value stored as a number only makes sense if we also know which number corresponds to which gender; a date stored as a number of days only makes sense if we know the reference date to count from; monetary values, temperatures, weights, and distances all require a unit to be meaningful.

The information that provides the context for a data value is an example of **metadata**.

Other examples of metadata include information about how data values were collected, such as *where* data values were recorded and *who* recorded them.

How should we store this information about the data values?

The short answer is that each piece of metadata is just itself a data value, so, in terms of computer memory, each individual piece of metadata can be stored as either a number or as text.

The larger question is how to store the metadata so that it is somehow connected to the raw data values. Deciding what data format to use for a particular data set should also take into account whether and how effectively metadata can be included alongside the core data values.

Recap

A block of bytes in computer memory can be interpreted in many ways. A binary format specifies how each byte or set of bytes in a file should be interpreted.

Extremely large numbers cannot be stored in computer memory using standard representations, and there is a limit to the precision with which real numbers can be stored.

Binary formats tend to use up less memory and provide faster access to data compared to text-based formats.

It is necessary to have specific software to work with a binary format, so it can be more expensive and it can be harder to share data.

5.4 Spreadsheets

It is important to mention spreadsheets as a storage option because spreadsheets are very widely used. However, it is also important to make a distinction between spreadsheet *software*, such as Microsoft Excel, and a spreadsheet *format* that is used to store a spreadsheet in computer memory, such as a Microsoft Excel workbook.

Spreadsheet software can be a very useful tool for viewing and exploring data, but using a spreadsheet format as the primary storage format for a data set is often not the best option.

5.4.1 Spreadsheet formats

One problem with spreadsheet formats is that they are specific to a particular piece of software.

For many people, a spreadsheet means a Microsoft Excel workbook. Until recently, Excel workbooks used a proprietary binary format, which implied that the Microsoft Excel software was necessary to make use of an Excel workbook. In practice, the Excel binary format was decoded and implemented by a number of different projects and the specification of the Microsoft Excel binary format is now publicly available, so Excel spreadsheets, particularly simple spreadsheets that contain only data, can be opened by a number of different software programs.

The latest versions of Excel store workbooks in an XML-based format called Open Office XML (OOXML), which promises a greater potential for working with Excel spreadsheets using other software. However, there is considerable controversy over the format and, at the time of writing it is not clear whether Excel workbooks in this format will be useable with software other than Microsoft Office products.

Excel is by far the most common spreadsheet software, but many other spreadsheet programs exist, notably the Open Office Calc software, which is an open source alternative and stores spreadsheets in an XML-based open standard format called Open Document Format (ODF). This allows the data to be accessed by a wider variety of software.

However, even ODF is not ideal as a storage format for a research data set because spreadsheet formats contain not only the data that is stored in the spreadsheet, but also information about how to display the data, such as fonts and colors, borders and shading.

Another problem with storing a data set in a spreadsheet is that a lot of unnecessary additional information is stored in the file. For example, information about the borders and shading of cells is also stored in a spreadsheet file. The spreadsheet may also include formulas to calculate cell values, images (charts), and more. Storing a data set in a Microsoft Excel workbook format is almost as bad as writing computer code using Microsoft Word (see Section 2.4.1).

In these ways, a spreadsheet format is less appropriate than a plain text or binary data format because it contains information that is not relevant to the data set and because the data can only be accessed using specific software.

In some spreadsheet formats, there are also limits on the numbers of columns and rows, so very large data sets simply cannot be accommodated.

5.4.2 Spreadsheet software

Spreadsheet software is useful because it displays a data set in a nice rectangular grid of cells. The arrangement of cells into distinct columns shares the same benefits as fixed-width format text files: it makes it very easy for a human to view and navigate within the data.

Most spreadsheet software also offers useful standard tools for manipulating the data. For example, it is easy to sort the data by a particular column. It is also possible to enter formulas into the cells of a spreadsheet so that, for example, sums and averages of columns may be obtained easily. In Excel, pivot tables (complex, interactive cross-tabulations) are a popular tool.

However, because most spreadsheet software provides facilities to import a data set from a wide variety of formats, these benefits of the spreadsheet *software* can be enjoyed without have to suffer the negatives of using a spreadsheet *format* for data storage. For example, it is possible to store the data set in a CSV format and use spreadsheet software to view or explore the data.

Furthermore, while spreadsheet software is very powerful and flexible, it is also quite lenient; it imposes very little discipline on the user. This tends to make it easy to introduce errors into a spreadsheet. We will see a more rigorous approach that provides powerful data manipulations in Section 5.6 on relational databases.

Finally, data exploration in spreadsheet software is typically conducted via menus and dialog boxes, which leaves no record of the steps taken. In chapter 9, we will look at writing computer code to explore data sets instead.

5.4.3 Case study: Over the limit

The Lada Riva is one of the highest-selling car models of all time and the only model to be found on every continent in the world.

A study conducted by researchers from the Psychology Department at the University of Auckland[3] looked at whether informative road signs had any effect on the speed at which vehicles travelled along a busy urban road in Auckland.

Data were collected for several days during a baseline period and for several days when each of five different signs were erected beside the road, for a total of six experimental stages. At each stage, the vehicle speeds were also collected for traffic travelling in the opposite direction along the road, to provide a set of "control" observations.

The data were collected by the Waitakere City Council via detectors buried in the road and were delivered to the researchers in the form of Excel spreadsheets. Figure 5.11 shows a section of one of these spreadsheets.

As we have discussed, spreadsheets are not an ideal data storage format, but this is an example where the researchers had no control over the format in which data are provided. This is why it is important to have some knowledge of a variety of data storage formats.

This figure demonstrates the fact that spreadsheet software displays the spreadsheet cells in a rectangular grid, which is very convenient for viewing the raw values in the data set. While it is not a replacement for proper data validation, taking a look at the raw values within a data set is never a bad thing.

Figure 5.11 also shows that it is straightforward to include metadata in a spreadsheet (cells A1 to A3) because each cell in the spreadsheet can contain any sort of value. On the other hand, just like in plain text formats, there is no way to indicate the special role of these cells; nothing in the file indicates where the real data begin.

Spreadsheets inherently provide a rows-and-columns view of the data, which, as we saw with plain text files, is not the most efficient way to represent data with a hierarchical structure.

[3]Wrapson, W., Harré, N, Murrell, P. (2006) Reductions in driver speed using posted feedback of speeding information: Social comparison or implied surveillance? *Accident Analysis and Prevention.* **38**, 1119–1126.

	A	B	C	D	E	F	G	H	I	J	K
1	Parrs Cross Road from Seymour Road, Daily Speed										
2	Monday 13/03/00										
3	Speed (KPH)										
4	Hour End	0 – 30	30 – 40	40 – 50	50 – 60	60 – 70	70 – 80	80 – 90	90–100	100–110	110–200
5											
6	1:00	0	1	14	26	15	2	0	0	0	0
7	2:00	0	1	7	13	5	2	0	0	0	0
8	3:00	0	0	1	2	5	1	2	0	0	0
9	4:00	0	0	3	2	2	2	1	0	0	0
10	5:00	0	0	5	4	2	0	1	0	0	0
11	6:00	0	2	8	28	17	3	0	0	0	0
12	7:00	0	4	45	110	39	3	1	0	0	0
13	7:15	2	2	37	78	17	1	0	0	0	0
14	7:30	0	2	65	84	26	0	0	0	0	0
15	7:45	1	0	53	160	16	0	0	0	0	0
16	8:00	2	13	43	125	45	2	0	0	0	0
17	7– 8	5	17	198	447	104	3	0	0	0	0
18	8:15	0	20	44	151	35	1	0	0	0	0
19	8:30	0	3	69	154	28	0	0	0	0	0
20	8:45	3	4	81	164	12	0	0	0	0	0
21	9:00	2	7	106	160	9	0	0	0	0	0
22	8– 9	5	34	300	629	84	1	0	0	0	0
23	10:00	0	12	225	460	80	6	0	0	0	1
24	11:00	3	13	128	313	68	4	1	2	0	0
25	12:00	6	18	180	353	62	1	0	0	0	1
26	13:00	6	11	133	383	84	3	0	0	1	0
27	14:00	12	16	196	329	55	3	0	0	0	0
28	15:00	5	28	156	351	74	1	0	0	0	0

Figure 5.11: Part of the vehicle speed data, as it was delivered, in a spreadsheet format.

In this case, the rows of the spreadsheet represent different times of the day and the columns represent different speed ranges. The actual data values are counts of the number of vehicles travelling within each speed range, within each time period. The spreadsheet shown in Figure 5.11 represents data from a single day of the study on one side of the road. How are the data for other days and data from the other side of the road stored?

Another nice feature of most spreadsheet formats is that they allow for multiple sheets within a document. This effectively provides a 3-dimensional cube of data cells, rather than just a 2-dimensional table of cells. The vehicle speed study made use of this feature to store the data from each day on a separate sheet, so that all of the data from one stage of the experiment and on one side of the road are stored in a single spreadsheet.

Figure 5.12 shows three sheets, representing three days' worth of data, within one of the spreadsheet files.

However, each experimental stage of the study was stored in a separate spreadsheet file, for a total of 12 spreadsheets. Once a data set becomes split across multiple files like this, there are issues of ensuring that all files retain the same structure, and there are issues with expressing the relationships between the separate files. For example, how do we know which files relate

	A	B	C	D	E	F	G	H	I	J	K
1	Parrs Cross Road from Seymour Road, Daily Speed										
2	Monday 13/03/00										
3	Speed (KPH)										
4	Hour End	0 – 30	30 – 40	40 – 50	50 – 60	60 – 70	70 – 80	80 – 90	90–100	100–110	110–200
5											
6	1:00										
7	2:00										
8	3:00										
9	4:00										
10	5:00										
11	6:00										
12	7:00										
13	7:15										
14	7:30										
15	7:45										
16	8:00										
17	7– 8										
18	8:15										
19	8:30										
20	8:45										
21	9:00										
22	8– 9										
23	10:										
24	11:										
25	12:										
26	13:										
27	14:										
28	15:										

	A	B	C	D	E	F	G	H	I	J	K
1	Parrs Cross Road from Seymour Road, Daily Speed										
2	Tuesday 14/03/00										
3	Speed (KPH)										
4	Hour End	0 – 30	30 – 40	40 – 50	50 – 60	60 – 70	70 – 80	80 – 90	90–100	100–110	110–200
5											
6	1:00										
7	2:00										
8	3:00										
9	4:00										
10	5:00										
11	6:00										
12	7:00										
13	7:15										
14	7:30										
15	7:45										
16	8:00										
17	7– 8										
18	8:15										
19	8:30										
20	8:45										
21	9:00										
22	8– 9										
23	10:										
24	11:										
25	12:										
26	13:										
27	14:										
28	15:										

	A	B	C	D	E	F	G	H	I	J	K
1	Parrs Cross Road from Seymour Road, Daily Speed										
2	Wednesday 15/03/00										
3	Speed (KPH)										
4	Hour End	0 – 30	30 – 40	40 – 50	50 – 60	60 – 70	70 – 80	80 – 90	90–100	100–110	110–200
5											
6	1:00	1	0	17	42	11	1	0	1	0	0
7	2:00	0	1	12	14	5	0	1	0	0	0
8	3:00	0	1	2	11	0	1	0	0	0	0
9	4:00	0	0	3	5	3	0	1	0	0	0
10	5:00	0	1	7	6	6	1	0	0	0	0
11	6:00	1	3	7	28	17	2	1	0	0	0
12	7:00	0	2	55	124	25	3	0	0	0	0
13	7:15	0	0	51	89	13	0	0	1	0	0
14	7:30	0	2	77	116	8	0	0	0	0	0
15	7:45	0	6	113	118	7	0	0	0	0	0
16	8:00	1	15	72	130	17	0	0	0	0	0
17	7– 8	1	23	313	453	45	0	0	1	0	0
18	8:15	0	21	117	130	5	0	0	0	0	0
19	8:30	0	8	121	117	11	3	0	0	0	0
20	8:45	19	42	122	74	9	0	0	0	0	0
21	9:00	56	64	118	82	5	0	0	0	0	0
22	8– 9	75	135	478	403	30	3	0	0	0	0
23	10:00	70	30	304	364	35	1	0	0	0	0
24	11:00	4	45	228	305	33	1	0	0	0	0
25	12:00	5	29	210	346	35	0	0	0	0	0
26	13:00	2	9	227	360	72	6	0	0	0	0
27	14:00	2	11	128	402	94	11	0	0	0	0
28	15:00	5	40	221	370	46	1	0	0	0	0

Figure 5.12: Three of the sheets in the vehicle speed data spreadsheet. Each sheet records speed data for one side of the road, for one day of the experiment.

to the same side of the road?

In this case, the names of the files can be used to differentiate between the side of the road and the experimental stage, but this naming scheme needs to be documented and explained somewhere. The problem is that there is no formal support for the person storing the data to express the structure of the data set. We will see better ways to resolve this problem in the next section on relational databases.

Another important problem is that spreadsheet cells are able to act independently of each other. Although the data in Figure 5.11 appear to have a useful structure, with each data value clearly associated with a time period and a speed range, this is in fact an illusion. The spreadsheet software does not, by default, place any significance on the fact that the values in row 6 all correspond to the time period from midnight to 1:00 a.m. Every cell in the spreadsheet is free to take any value regardless of which row or column it resides in.

This problem can be seen by looking at the time values in column A. To the human observer, it is clear that this column of values (apart from the first three rows) corresponds to time intervals. However, the spreadsheet data model does not enforce any such constraint on the data, as the value in row 17 clearly shows. All of the values up to that point (rows 6 through 16) have been time values, but row 17 contains the value 7-8. To human eyes this is clearly the time period 7:00 a.m. to 8:00 a.m., but any software trying to read this column of values will almost certainly fail to make that intuitive leap.

This particular problem is a feature of this particular data set, but the general problem pertaining to all spreadsheets is that the flexible value-per-cell data model allows this sort of thing to happen and the consequence is that additional data cleaning is necessary before the raw data in the spreadsheet can be used for analysis.

It is a little unfair to point out these problems with a spreadsheet format example because most of these problems also exist for plain text files. The main point is that the spreadsheet *format* does not provide a great deal beyond much simpler format options *and* it introduces new problems of its own.

Recap

Spreadsheet software is a useful tool for viewing and exploring data.

A spreadsheet storage format does allow for a 3-dimensional cube of data rather than just a 2-dimensional table of data, but it requires specific software and is less efficient because it stores extraneous information.

5.5 XML

One of the main problems with plain text formats is that there is no information within the file itself to describe the location of the data values.

One solution to this problem is to provide a recognizable label for each data value within the file.

This section looks at the **eXtensible Markup Language**, XML, which provides a formal way to provide this sort of labeling or "markup" for data.

Data stored in an XML document is stored as characters, just like in a plain text format, but the information is organized within the file in a much more sophisticated manner.

As a simple example to get us started, Figure 5.13 shows the surface temperature data at the Pacific Pole of Inaccessibility (see Section 1.1) in two different formats: an XML format and the original plain text format.

The major difference between these two storage formats is that, in the XML version, every single data value is distinctly labeled. This means that, for example, even without any background explanation of this data set, we could easily identify the "temperature" values within the XML file. By contrast, the same task would be impossible with the plain text format, unless additional information is provided about the data set and the arrangement of the values within the file. Importantly, the XML format allows a computer to perform this task completely automatically because it could detect the values with a label of `temperature` within the XML file.

One fundamental *similarity* between these formats is that they are both just text. This is an important and beneficial property of XML; we can read it and manipulate it without any special skills or any specialized software.

XML is a storage format that is still based on plain text but does not suffer from many of the problems of plain text files because it adds flexibility, rigor, and standardization.

Many of the benefits of XML arise from the fact that it is a computer language. When we store a data set in XML format, we are writing computer code that expresses the data values to the computer. This is a huge step beyond free-form plain text files because we are able to communicate much more about the data set to the computer. However, the cost is that we need to learn the rules of XML so that we can communicate the data correctly.

In the next few sections, we will focus on the details of XML. In Section 5.5.1, we will look briefly at the syntax rules of XML, which will allow us to store data *correctly* in an XML format. We will see that there is a great degree of choice when using XML to store a data set, so in Sections 5.5.2 and 5.5.3, we will discuss some ideas about how to store data *sensibly* in an XML format—some uses of XML are better than others. In Section 5.5.5 we will return to a more general discussion of how XML formats compare to other storage formats.

The information in this section will be useful whether we are called upon to create an XML document ourselves or whether we just have to work with a data set that someone else has chosen to store in an XML format.

```
<?xml version="1.0"?>
<temperatures>
    <variable>Mean TS from clear sky composite (kelvin)</variable>
    <filename>ISCCPMonthly_avg.nc</filename>
    <filepath>/usr/local/fer_dsets/data/</filepath>
    <subset>48 points (TIME)</subset>
    <longitude>123.8W(-123.8)</longitude>
    <latitude>48.8S</latitude>
    <case date="16-JAN-1994" temperature="278.9" />
    <case date="16-FEB-1994" temperature="280" />
    <case date="16-MAR-1994" temperature="278.9" />
    <case date="16-APR-1994" temperature="278.9" />
    <case date="16-MAY-1994" temperature="277.8" />
    <case date="16-JUN-1994" temperature="276.1" />

    ...

</temperatures>
```

```
             VARIABLE : Mean TS from clear sky composite (kelvin)
             FILENAME : ISCCPMonthly_avg.nc
             FILEPATH : /usr/local/fer_data/data/
             SUBSET   : 48 points (TIME)
             LONGITUDE: 123.8W(-123.8)
             LATITUDE : 48.8S
                       123.8W
                        23
 16-JAN-1994 00 /  1:  278.9
 16-FEB-1994 00 /  2:  280.0
 16-MAR-1994 00 /  3:  278.9
 16-APR-1994 00 /  4:  278.9
 16-MAY-1994 00 /  5:  277.8
 16-JUN-1994 00 /  6:  276.1
 ...
```

Figure 5.13: The first few lines of the surface temperature at Point Nemo in two formats: an XML format (top) and the original plain text format (bottom).

5.5.1 XML syntax

As we saw in Chapter 2, the first thing we need to know about a computer language are the syntax rules, so that we can write code that is correct.

We will use the XML format for the Point Nemo temperature data (Figure 5.13) to demonstrate some of the basic rules of XML syntax.

The XML format of the data consists of two parts: XML **markup** and the actual data itself. For example, the information about the latitude at which these data were recorded is stored within XML **tags**, `<latitude>` and `</latitude>`. The combination of tags and content is together described as an XML **element**.

```
<latitude>48.8S</latitude>
```

```
   element:  <latitude>48.8S</latitude>
 start tag:  <latitude>48.8S</latitude>
data value:  <latitude>48.8S</latitude>
   end tag:  <latitude>48.8S</latitude>
```

Each temperature measurement in the XML file is contained within a `case` element, with the date and temperature data recorded as **attributes** of the element. The values of attributes must be contained within double-quotes.

```
<case date="16-JAN-1994"
      temperature="278.9" />
```

```
   element name:  <case date="16-JAN-1994"
 attribute name:  <case date="16-JAN-1994"
attribute value:  <case date="16-JAN-1994"
 attribute name:        temperature="278.9" />
attribute value:        temperature="278.9" />
```

This should look very familiar because these are exactly the same notions of elements and attributes that we saw in HTML documents in Chapter 2. However, there are some important differences between XML syntax and HTML syntax. For example, XML is case-sensitive so the names in the start and end tags for an element must match exactly. Some other differences are detailed below.

XML document structure

The first line of an XML document must be a declaration that the file is an XML document and which version of XML is being used.

```
<?xml version="1.0"?>
```

The declaration can also include information about the encoding of the XML document (see Section 5.2.7), which is important for any software that needs to read the document.

```
<?xml version="1.0" encoding="UTF-8"?>
```

Unlike HTML, where there is a fixed set of allowed elements, an XML document can include elements with any name whatsoever, as long as the elements are all nested cleanly and there is only one outer element, called the **root** element, that contains all other elements in the document.

In the example XML document in Figure 5.13, there is a single `temperatures` element, with all other elements, e.g., `variable` and `case` elements, nested within that.

Elements can be empty, which means that they consist only of a start tag (no content and no end tag), but these empty elements must end with `/>`, rather than the normal `>`. The code below shows an example of an empty `case` element.

```
    <case date="16-JAN-1994" temperature="278.9" />
```

Although it is possible to use any element and attribute names in an XML document, in practice, data sets that are to be shared will adhere to an agreed-upon set of elements. We will discuss how an XML document can be restricted to a fixed set of elements later on in Section 5.5.3.

Code layout

Because XML is a computer language, there is a clear structure within the XML code. As discussed in Section 2.4.3, we should write XML code using techniques such as indenting so that the code is easy for people to read and understand.

Escape sequences

As with HTML, the characters <, >, and & (among others) are special and must be replaced with special escape sequences, <, >, and & respectively.

These escape sequences can be very inconvenient when storing data values, so it is also possible to mark an entire section of an XML document as "plain text" by placing it within a special CDATA section, within which all characters lose their special meaning.

As an example, shown below is one of the data values that we wanted to store for the clay tablet YBC 7289 in Section 5.1. These represent the markings that are etched into the clay tablet.

```
| <<|||| <<<<<| <
```

If we wanted to store this data within an XML document, we would have to escape every one of the < symbols. The following code shows what the data would look like within an XML document.

```
| &lt;&lt;|||| &lt;&lt;&lt;&lt;&lt;| &lt;
```

The special CDATA syntax allows us to avoid having to use these escape sequences. The following code shows what the XML code would look like within an XML document within a CDATA section.

```
<![CDATA[
  | <<|||| <<<<<| <
]]>
```

Checking XML syntax

There are many software packages that can read XML and most of these will report any problems with XML syntax.

The W3C Markup Validation Service[4] can be used to perform an explicit check on an XML document online. Alternatively, the libxml software library[5] can be installed on a local computer. This software includes a command-line tool called xmllint for checking XML syntax. For simple use of xmllint, the only thing we need to know is the name of the XML

[4] http://validator.w3.org/

[5] http://xmlsoft.org/

document and where that file is located. Given an XML document called `xmlcode.xml`, the following command could be entered in a command window or terminal to check the syntax of the XML code in the file.

```
xmllint xmlcode.xml
```

There is no general software that "runs" XML code because the code does not really "do" anything. The XML markup is just there to describe a set of data. However, we will see how to retrieve the data from an XML document later in Sections 7.3.1 and 9.7.7.

5.5.2 XML design

Although there are syntax rules that any XML document must follow, we still have a great deal of flexibility in how we choose to mark up a particular data set. It is possible to use *any* element names and *any* attribute names that we like.

For example, in Figure 5.13, each set of measurements, a date plus a temperature reading, is stored within a `case` element, using `date` and `temperature` attributes. The first set of measurements in this format is repeated below.

```
<case date="16-JAN-1994" temperature="278.9" />
```

It would still be correct syntax to store these measurements using different element and attribute names, as in the code below.

```
<record time="16-JAN-1994" temp="278.9" />
```

It is important to choose names for elements and attributes that are meaningful, but in terms of XML syntax, there is no single best answer.

When we store a data set as XML, we have to decide which elements and attributes to use to store the data. We also have to consider how elements nest within each other, if at all. In other words, we have to decide upon a *design* for the XML document.

In this section, we will look at some of these questions and consider some solutions for how to design an XML document.

Even if, in practice, we are not often faced with the prospect of designing an XML format, this discussion will be useful in understanding why an XML document that we encounter has a particular structure. It will also be useful as an introduction to similar design ideas that we will discuss when we get

```
<?xml version="1.0"?>
<temperatures>
    <variable>Mean TS from clear sky composite (kelvin)</variable>
    <filename>ISCCPMonthly_avg.nc</filename>
    <filepath>/usr/local/fer_dsets/data/</filepath>
    <subset>93 points (TIME)</subset>
    <longitude>123.8W(-123.8)</longitude>
    <latitude>48.8S</latitude>
    <cases>
16-JAN-1994 278.9
16-FEB-1994 280.0
16-MAR-1994 278.9
16-APR-1994 278.9
16-MAY-1994 277.8
16-JUN-1994 276.1

    ...

    </cases>
</temperatures>
```

Figure 5.14: The first few lines of the surface temperature at Point Nemo in an alternative XML format. This format should be compared with the XML format in Figure 5.13.

to relational databases in Section 5.6.

Marking up data

The first XML design issue is to make sure that each value within a data set can be clearly identified. In other words, it should be trivial for a computer to extract each individual value. This means that every single value should be either the content of an element or the value of an attribute. The XML document shown in Figure 5.13 demonstrates this idea.

Figure 5.14 shows another possible XML representation of the Pacific Pole of Inaccessibility temperature data.

In this design, the irregular and one-off metadata values are individually identified within elements or attributes, but the regular and repetitive raw data values are not. This is not ideal from the point of view of labeling every individual data value within the file, but it may be a viable option when the raw data values have a very simple format (e.g., comma-delimited) and the data set is very large, in which case avoiding lengthy tags and attribute names would be a major saving.

The main point is that, as with the selection of element and attribute names, there is no single best markup strategy for all possible situations.

Things and measurements on things

When presented with a data set, the following questions should guide the design of the XML format:

Things that have been measured:
What sorts of objects or "things" have been measured? These typically correspond not only to the subjects in an experiment or survey, but also to any groupings of subjects, for example, families, neighborhoods, or plots in a field.

Measurements that have been made:
What measurements have been made on each object or "thing"? These correspond to the traditional idea of responses or *dependent* variables.

A simple rule of thumb is then to have an element for each object in the data set (and a different *type* of element for each different type of object) and then have an attribute for each measurement in the data set.

For example, consider the family tree data set in Figure 5.15 (this is a reproduction of Figure 5.3 for convenience). This data set contains demographic information about several related people.

In this case, there are obviously measurements taken on people, those measurements being names, ages, and genders. We could distinguish between parent objects and child objects, so we have elements for each of these.

```
<parent gender="female" name="Julia" age="32" />
<child gender="male" name="Jack" age="6" />
```

When a data set has a hierarchical structure, an XML document can be designed to store the information more efficiently and more appropriately by nesting elements to avoid repeating values.

For example, there are two distinct families of people, so we could have elements to represent the different families and nest the relevant people within the appropriate family element to represent membership of a family.

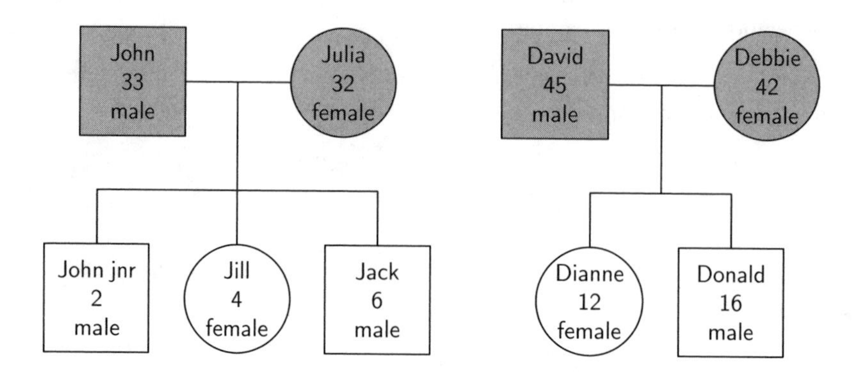

Figure 5.15: An example of hierarchical data: a family tree containing data on parents (grey) and children (white). This figure is a reproduction of Figure 5.3.

```
<family>
  <parent gender="male" name="John" age="33" />
  <parent gender="female" name="Julia" age="32" />
  <child gender="male" name="Jack" age="6" />
  <child gender="female" name="Jill" age="4" />
  <child gender="male" name="John jnr" age="2" />
</family>
<family>
  <parent gender="male" name="David" age="45" />
  <parent gender="female" name="Debbie" age="42" />
  <child gender="male" name="Donald" age="16" />
  <child gender="female" name="Dianne" age="12" />
</family>
```

Elements versus attributes

Another decision to make is whether to store data values as the values of attributes of XML elements or as the content of XML elements. For example, when storing the Point Nemo temperature data, we could store the temperature values as attributes of a `case` element as shown in Figure 5.13 and below.

```
<case temperature="278.9" />
```

Alternatively, we could store the temperature values as the content of a `temperature` element.

```
<temperature>278.9</temperature>
```

As demonstrated so far, one simple solution is to store all measurements as the values of attributes. Besides its simplicity, this approach also has the advantage of providing a simple translation between the XML format and a corresponding relational database design. We will return to this point in Section 5.6.7.

However, there is considerable controversy on this point. One standard viewpoint is that the content of elements is data and the attributes are metadata. Another point *against* storing data in attributes is that it is not always possible or appropriate. For example, a data value may have to be stored as the content of a separate element, rather than as the value of an attribute in the following cases:

Large values
When the data value contains a lot of information, such as a general comment consisting of paragraphs of text, the value is better stored as the content of an element.

Special characters
When the measurement contains lots of special characters, which would require a lot of escape sequences, the value can be stored much more easily as the content of an element.

Order of data values
When the order of the measurements matters, storing values as the content of elements is more appropriate because the order of attributes is arbitrary, but the order of elements within a document matters.

Complex values
When the measurements are not "simple" values—in other words, when a measurement is actually a series of measurements on a different sort of object (e.g., information about a room within a building)—the values are better stored as the content of an element. This is another way of saying that the value of an attribute cannot be an XML element. Data values that are not atomic (a single value) will generate an entire XML element, which must be stored as the content of a parent XML element.

5.5.3 XML schema

Having settled on a particular design for an XML document, we need to be able to write down the design in some way.

We need the design written down so that we can check that an XML document follows the design that we have chosen *and* so that we can communicate our design to others so that we can share XML documents that have this particular format. In particular, we need to write the design down in such a way that a computer understands the design, so that a computer can check that an XML document obeys the design.

Yes, we need to learn *another* computer language.

The way that the XML design can be specified is by creating a **schema** for an XML document, which is a description of the structure of the document. A number of technologies exist for specifying XML schema, but we will focus on the **Document Type Definition** (DTD) language.

A DTD is a set of rules for an XML document. It contains **element declarations** that describe which elements are permitted within the XML document, in what order, and how they may be nested within each other. The DTD also contains **attribute declarations** that describe what attributes an element can have, whether attributes are optional or not, and what sort of values each attribute can have.

5.5.4 Case study: Point Nemo (continued)

Figure 5.16 shows the temperature data at Point Nemo in an XML format (this is a reproduction of part of Figure 5.13 for convenience).

The structure of this XML document is as follows. There is a single overall `temperatures` element that contains all other elements. There are several elements containing various sorts of metadata: a `variable` element containing a description of the variable that has been measured; a `filename` element and a `filepath` element containing information about the file from which these data were extracted; and three elements, `subset`, `longitude`, and `latitude`, that together describe the temporal and spatial limits of this data set. Finally, there are a number of `case` elements that contain the core temperature data; each `case` element contains a temperature measurement and the date of the measurement as attributes.

A DTD describing this structure is shown in Figure 5.17.

The DTD code consists of two types of declarations. There must be an

```
<?xml version="1.0"?>
<temperatures>
    <variable>Mean TS from clear sky composite (kelvin)</variable>
    <filename>ISCCPMonthly_avg.nc</filename>
    <filepath>/usr/local/fer_dsets/data/</filepath>
    <subset>93 points (TIME)</subset>
    <longitude>123.8W(-123.8)</longitude>
    <latitude>48.8S</latitude>
    <case date="16-JAN-1994" temperature="278.9" />
    <case date="16-FEB-1994" temperature="280" />
    <case date="16-MAR-1994" temperature="278.9" />
    <case date="16-APR-1994" temperature="278.9" />
    <case date="16-MAY-1994" temperature="277.8" />
    <case date="16-JUN-1994" temperature="276.1" />

    ...

</temperatures>
```

Figure 5.16: The first few lines of the surface temperature at Point Nemo in an XML format. This is a reproduction of part of Figure 5.13.

```
 1 <!ELEMENT temperatures (variable,
 2                         filename,
 3                         filepath,
 4                         subset,
 5                         longitude,
 6                         latitude,
 7                         case*)>
 8 <!ELEMENT variable (#PCDATA)>
 9 <!ELEMENT filename (#PCDATA)>
10 <!ELEMENT filepath (#PCDATA)>
11 <!ELEMENT subset (#PCDATA)>
12 <!ELEMENT longitude (#PCDATA)>
13 <!ELEMENT latitude (#PCDATA)>
14 <!ELEMENT case EMPTY>
15
16 <!ATTLIST case
17     date        ID    #REQUIRED
18     temperature CDATA #IMPLIED>
```

Figure 5.17: A DTD for the XML format used to store the surface temperature at Point Nemo (see Figure 5.16). The line numbers (in grey) are just for reference.

<!ELEMENT> declaration for each type of element that appears in the XML design and there must be an <!ATTLIST> declaration for every element in the design that has one or more attributes.

The main purpose of the <!ELEMENT> declarations is to specify what is allowed as the content of a particular type of element. The simplest example of an <!ELEMENT> declaration is for case elements (line 14) because they are empty (they have no content), as indicated by the keyword EMPTY. The components of this declaration are shown below.

```
     keyword:  <!ELEMENT case EMPTY>
element name:  <!ELEMENT case EMPTY>
     keyword:  <!ELEMENT case EMPTY>
```

The keywords ELEMENT and EMPTY will be the same for the declaration of any empty element. All that will change is the name of the element.

Most other elements are similarly straightforward because their contents are just text, as indicated by the #PCDATA keyword (lines 8 to 13). These examples demonstrate that the declaration of the content of the element is specified within parentheses. The components of the declaration for the longitude element are shown below.

```
        keyword:  <!ELEMENT longitude (#PCDATA)>
   element name:  <!ELEMENT longitude (#PCDATA)>
    parentheses:  <!ELEMENT longitude (#PCDATA)>
element content:  <!ELEMENT longitude (#PCDATA)>
```

The temperatures element is more complex because it can contain other elements. The declaration given in Figure 5.17 (lines 1 to 7) specifies seven elements (separated by commas) that are allowed to be nested within a temperatures element. The order of these elements within the declaration is significant because this order is imposed on the elements in the XML document. The first six elements, variable to latitude, are compulsory because there are no modifiers after the element names; exactly one of each element must occur in the XML document. The case element, by contrast, has an asterisk, *, after it, which means that there can be zero or more case elements in the XML document.

The purpose of the <!ATTLIST> declarations in a DTD is to specify which attributes each element is allowed to have. In this example, only the case elements have attributes, so there is only one <!ATTLIST> declaration (lines 16 to 18). This declaration specifies three things for each attribute: the name of the attribute, what sort of value the attribute can have, and whether the attribute is compulsory or optional. The components of this declaration are shown below.

keyword:	<!ATTLIST case		
element name:	<!ATTLIST case		
attribute name:	date	ID	#REQUIRED
attribute value:	date	ID	#REQUIRED
compulsory attribute:	date	ID	#REQUIRED
attribute name:	temperature	CDATA	#IMPLIED>
attribute value:	temperature	CDATA	#IMPLIED>
optional attribute:	temperature	CDATA	#IMPLIED>

The date attribute for case elements is compulsory (#REQUIRED) and the value must be unique (ID). The temperature attribute is optional (#IMPLIED) and, if it occurs, the value can be any text (CDATA).

Section 6.2 describes the syntax and semantics of DTD files in more detail.

The rules given in a DTD are associated with an XML document by adding a **Document Type Declaration** as the second line of the XML document. This can have one of two forms:

DTD inline:

The DTD can be included within the XML document. In the Point Nemo example, it would look like this:

```
<?xml version="1.0"?>
<!DOCTYPE temperatures [
    DTD code
 ]>
<temperatures>
    ...
```

External DTD

The DTD can be in an external file, say pointnemotemp.dtd, and the XML document can refer to that file:

```
<?xml version="1.0"?>
<!DOCTYPE temperatures SYSTEM "pointnemotemp.dtd">
<temperatures>
    ...
```

The DRY principle suggests that an external DTD is the most sensible approach because then the same DTD rules can be applied efficiently to many XML documents.

An XML document is said to be **well-formed** if it obeys the basic rules of XML syntax. If the XML document *also* obeys the rules given in a DTD, then the document is said to be **valid**. A valid XML document has the advantage that we can be sure that all of the necessary information for a data set has been included and has the correct structure, and that all data values have the correct sort of value.

The use of a DTD has some shortcomings, such as a lack of support for precisely specifying the data type of attribute values or the contents of elements. For example, it is not possible to specify that an attribute value must be an integer value between 0 and 100. There is also the difficulty that the DTD language is completely different from XML, so there is another technology to learn. XML Schema is an XML-based technology for specifying the design of XML documents that solves both of those problems, but it comes at the cost of much greater complexity. This complexity has led to the development of further technologies that simplify the XML Schema syntax, such as Relax NG.

Standard schema

So far we have discussed designing our own XML schema to store data in an XML document. However, many standard XML schema already exist, so another option is simply to choose one of those instead and create an XML document that conforms to the appropriate standard.

These standards have typically arisen in a particular area of research or business. For example, the Statistical Data and Metadata eXchange (SDMX) format has been developed by several large financial institutions for sharing financial data, and the Data Documentation Initiative (DDI) is aimed at storing metadata for social science data sets.

One downside is that these standards can be quite complex and may require expert assistance and specialized software to work with the appropriate format, but the upside is integration with a larger community of researchers and compatibility with a wider variety of software tools.

5.5.5 Advantages and disadvantages

We will now consider XML not just as an end in itself, but as one of many possible storage formats. In what ways is the XML format better or worse than other storage options, particularly the typical unstructured plain text format that we saw in Section 5.2?

A self-describing format

The core advantage of an XML document is that it is **self-describing**.

The tags in an XML document provide information about where the data is stored within the document. This is an advantage because it means that humans can find information within the file easily. That is true of any plain text file, but it is especially true of XML files because the tags essentially provide a level of documentation for the human reader. For example, the XML element shown below not only makes it easy to determine that the value `48.8S` constitutes a single data value within the file, but it also makes it clear that this value is a north–south geographic location.

```
<latitude>48.8S</latitude>
```

The fact that an XML document is self-describing is an even greater advantage from the perspective of the computer. An XML document provides enough information for software to determine how to read the file, without any further human intervention. Look again at the line containing latitude information.

```
<latitude>48.8S</latitude>
```

There is enough information for the computer to be able to detect the value `48.8S` as a single data value, and the computer can also record the `latitude` label so that if a human user requests the information on latitude, the computer knows what to provide.

One consequence of this feature that may not be immediately obvious is that it is much easier to modify the structure of data within an XML document compared to a plain text file. The location of information within an XML document is not so much dependent on where it occurs within the file, but where the tags occur within the file. As a trivial example, consider reversing the order of the following lines in the Point Nemo XML file.

```
<longitude>123.8W(-123.8)</longitude>
<latitude>48.8S</latitude>
```

If the information were stored in the reverse order, as shown below, the task of retrieving the information on latitude would be exactly the same. This can be a huge advantage if larger modifications need to be made to a data set, such as adding an entire new variable.

```
<latitude>48.8S</latitude>
<longitude>123.8W(-123.8)</longitude>
```

Representing complex data structures

The second main advantage of the XML format is that it can accommodate complex data structures. Consider the hierarchical data set in Figure 5.15. Because XML elements can be nested within each other, this sort of data set can be stored in a sensible fashion with families grouped together to make parent–child relations implicit and avoid repetition of the parent data. The plain text representation of these data are reproduced from page 74 below along with a possible XML representation.

```
                   John      33  male
                   Julia     32  female
John    Julia      Jack       6  male
John    Julia      Jill       4  female
John    Julia      John jnr   2  male
                   David     45  male
                   Debbie    42  female
David   Debbie     Donald    16  male
David   Debbie     Dianne    12  female
```

```
<family>
  <parent gender="male" name="John" age="33" />
  <parent gender="female" name="Julia" age="32" />
  <child gender="male" name="Jack" age="6" />
  <child gender="female" name="Jill" age="4" />
  <child gender="male" name="John jnr" age="2" />
</family>
<family>
  <parent gender="male" name="David" age="45" />
  <parent gender="female" name="Debbie" age="42" />
  <child gender="male" name="Donald" age="16" />
  <child gender="female" name="Dianne" age="12" />
</family>
```

The XML format is superior in the sense that the information about each person is only recorded once. Another advantage is that it would be very easy to represent a wider range of situations using the XML format. For example, if we wanted to allow for a family unit to have a third parent (e.g., a step-parent), that would be straightforward in XML, but it would be much more awkward in the fixed rows-and-columns plain text format.

Data integrity

Another important advantage of the XML format is that it provides some level of checking on the correctness of the data file (a check on the **data integrity**). First of all, there is the fact that any XML document must obey the rules of XML, which means that we can use a computer to check that an XML document at least has a sensible structure.

If an XML document also has a DTD, then we can perform much more rigid checks on the correctness of the document. If data values are stored as attribute values, it is possible for the DTD to provide checks that the data values themselves are valid. The XML Schema language provides even greater facilities for specifying limits and ranges on data values.

Verbosity

The major disadvantage of XML is that it generates large files. With its being a plain text format, it is not memory efficient to start with, and with all of the additional tags around the actual data, files can become extremely large. In many cases, the tags can take up more room than the actual data!

These issues can be particularly acute for research data sets, where the structure of the data may be quite straightforward. For example, geographic data sets containing many observations at fixed locations naturally form a 3-dimensional array of values, which can be represented very simply and efficiently in a plain text or binary format. In such cases, having highly repetitive XML tags around all values can be very inefficient indeed.

The verbosity of XML is also a problem for entering data into an XML format. It is just too laborious for a human to enter all of the tags by hand, so, in practice, it is only sensible to have a computer generate XML documents.

Costs of complexity

It should also be acknowledged that the additional sophistication of XML creates additional costs. Users have to be more educated and the software has to be more complex, which means that fewer software packages are able to cope with data stored as XML.

In summary, the fact that computers can read XML easily and effectively, plus the fact that computers can produce XML rapidly (verbosity is less of an issue for a computer), means that XML is an excellent format for transferring information between different software programs. XML is a good language for computers to use to talk to each other, with the added bonus that humans can still easily eavesdrop on the conversation.

Recap

XML is a language for describing a data set.

XML consists of elements and attributes, with data values stored as the content of elements or as the values of attributes.

The design of an XML document—the choice of elements and attributes—is important. One approach has an element for each different object that has been measured, with the actual measurements recorded as attributes of the appropriate element.

The DTD language can be used to formally describe the design of an XML document.

The major advantage of XML is that XML documents are self-describing, which means that each data value is unambiguously labeled within the file, so that a computer can find data values without requiring additional information about the file.

5.6 Databases

When a data set becomes very large, or even just very complex in its structure, the ultimate storage solution is a **database**.

The term "database" can be used generally to describe any collection of information. In this section, the term "database" means a **relational database**, which is a collection of data that is organized in a particular way.

The actual physical storage mechanism for a database—whether binary formats or text formats are used, and whether one file or many files are used—will not concern us. We will only be concerned with the high-level, conceptual organization of the data and will rely on software to decide how

ISBN	title	author	gender	publisher	ctry
0618260307	The Hobbit	J. R. R. Tolkien	male	Houghton Mifflin	USA
0908606664	Slinky Malinki	Lynley Dodd	female	Mallinson Rendel	NZ
1908606206	Hairy Maclary from Donaldson's Dairy	Lynley Dodd	female	Mallinson Rendel	NZ
0393310728	How to Lie with Statistics	Darrell Huff	male	W. W. Norton	USA
0908783116	Mechanical Harry	Bob Kerr	male	Mallinson Rendel	NZ
0908606273	My Cat Likes to Hide in Boxes	Lynley Dodd	female	Mallinson Rendel	NZ
0908606273	My Cat Likes to Hide in Boxes	Eve Sutton	female	Mallinson Rendel	NZ

Figure 5.18: Information about a set of books, including the ISBN and title for the book, the author of the book and the author's gender, the publisher of the book, and the publisher's country of origin.

best to store the information in files.

The software that handles the representation of the data in computer memory, and allows us to work at a conceptual level, is called a **database management system** (**DBMS**), or in our case, more specifically, a *relational* database management system (**RDBMS**).

The main benefits of databases for data storage derive from the fact that databases have a formal structure. We will spend much of this section describing and discussing how databases are designed, so that we can appreciate the benefits of storing data in a database and so that we know enough to be able to work with data that have been stored in a database.

5.6.1 The database data model

We are not concerned with the file formats that are used to store a database. Instead, we will deal with the conceptual components used to store data in a database.

A relational database consists of a set of **tables**, where a table is conceptually just like a plain text file or a spreadsheet: a set of values arranged in rows and columns. The difference is that there are usually several tables in a single database, and the tables in a database have a much more formal structure than a plain text file or a spreadsheet.

In order to demonstrate the concepts and terminology of databases, we will work with a simple example of storing information about books. The entire set of information is shown in Figure 5.18, but we will only consider specific subsets of book information at various stages throughout this section in order to demonstrate different ideas about databases.

Shown below is a simple example of a database table that contains information about some of the books in our data set. This table has three columns—the ISBN of the book, the title of the book, and the author of the book—and four rows, with each row representing one book.

ISBN	title	author
0618260307	The Hobbit	J. R. R. Tolkien
0908606664	Slinky Malinki	Lynley Dodd
0393310728	How to Lie with Statistics	Darrell Huff
0908783116	Mechanical Harry	Bob Kerr

Each table in a database has a unique name and each column within a table has a unique name within that table.

Each column in a database table also has a **data type** associated with it, so all values in a single column are the same sort of data. In the book database example, all values in all three columns are text or character values. The ISBN is stored as text, not as an integer, because it is a sequence of 10 digits (as opposed to a decimal value). For example, if we stored the ISBN as an integer, we would lose the leading `0`.

Each table in a database has a **primary key**. The primary key must be unique for every row in a table. In the book table, the ISBN provides a perfect primary key because every book has a different ISBN.

It is possible to create a primary key by combining the values of two or more columns. This is called a **composite primary key**. A table can only have one primary key, but the primary key may be composed from more than one column. We will see some examples of composite primary keys later in this chapter.

A database containing information on books might also contain information on book publishers. Below we show another table *in the same database* containing information on publishers.

ID	name	country
1	Mallinson Rendel	NZ
2	W. W. Norton	USA
3	Houghton Mifflin	USA

In this table, the values in the `ID` column are all integers. The other columns all contain text. The primary key in this table is the `ID` column.

Tables within the same database can be related to each other using **for-**

eign keys. These are columns in one table that specify a value from the primary key in another table. For example, we can relate each book in the `book_table` to a publisher in the `publisher_table` by adding a foreign key to the `book_table`. This foreign key consists of a column, `pub`, containing the appropriate publisher `ID`. The `book_table` now looks like this:

```
ISBN        title                         author            pub
----------  --------------------------    ----------------  ---
0618260307  The Hobbit                    J. R. R. Tolkien  3
0908606664  Slinky Malinki                Lynley Dodd       1
0393310728  How to Lie with Statistics    Darrell Huff      2
0908783116  Mechanical Harry              Bob Kerr          1
```

Notice that two of the books in the `book_table` have the same publisher, with a `pub` value of `1`. This corresponds to the publisher with an `ID` value of `1` in the `publisher_table`, which is the publisher called Mallinson Rendel.

Also notice that a foreign key column in one table does *not* have to have the same name as the primary key column that it refers to in another table. The foreign key column in the `book_table` is called `pub`, but it refers to the primary key column in the `publisher_table` called `ID`.

5.6.2 Database notation

The examples of database tables in the previous section have shown the *contents* of each database table. In the next section, on Database Design, it will be more important to describe the *design*, or structure, of a database table—the table **schema**. For this purpose, the contents of each row are not important; instead, we are most interested in how many tables there are and which columns are used to make up those tables.

We can describe a database design simply in terms of the names of tables, the names of columns, which columns are primary keys, and which columns are foreign keys.

The notation that we will use is a simple text description, with primary keys and foreign keys indicated in square brackets. The description of a foreign key includes the name of the table and the name of the column that the foreign key refers to. For example, these are the schema for the `publisher_table` and the `book_table` in the book database:

```
publisher_table  ( ID [PK], name, country )
```

```
book_table  ( ISBN [PK], title, author,
              pub [FK publisher_table.ID] )
```

The diagram below shows one way that this design could be visualized. Each "box" in this diagram represents one table in the database, with the name of the table as the heading in the box. The other names in each box are the names of the *columns* within the table; if the name is **bold**, then that column is part of the primary key for the table and if the name is *italic*, then that column is a foreign key. Arrows are used to show the link between a foreign key in one table and the primary key in another table.

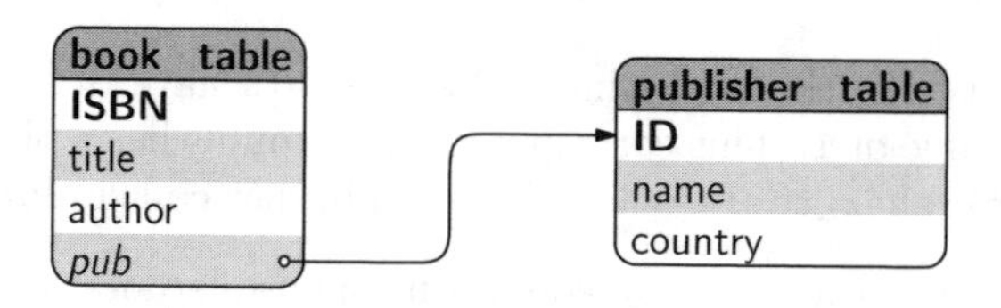

The `publisher_table` has three columns and the column named `ID` is the primary key for the table.

The `book_table` has four columns. In this table, the primary key is the `ISBN` column and the `pub` column is a foreign key that refers to the `ID` column in the `publisher_table`.

5.6.3 Database design

Like we saw with XML documents in Section 5.5.2, databases allow us to store information in a variety of ways, which means that there are design decisions to be made. In this section, we will briefly discuss some of the issues relating to database design.

The design of a database comes down to three things: how many tables are required; what information goes in each table; and how the tables are linked to each other. The remainder of this section provides some rules and guidelines for determining a solution for each of these steps.

This section provides neither an exhaustive discussion nor a completely rigorous discussion of database design. The importance of this section is to provide a basic introduction to some useful ideas and ways to think about data. A basic understanding of these issues is also necessary for us to be able to work with data that have been stored in a database.

Entities and attributes

One way to approach database design is to think in terms of entities, their attributes, and the relationships between them.

An **entity** is most easily thought of as a person, place, or physical object (e.g., a book); an event; or a concept. An **attribute** is a piece of information about the entity. For example, the title, author, and ISBN are all attributes of a book entity.

In terms of a research data set, each variable in the data set corresponds to an attribute. The task of designing a database to store the data set comes down to assigning each variable to a particular entity.

Having decided upon a set of entities and their attributes, a database design consists of a separate table for each entity and a separate column within each table for each attribute of the corresponding entity.

Rather than storing a data set as one big table of information, this rule suggests that we should use several tables, with information about different entities in separate tables. In the book database example, there is information about at least two entities, books and publishers, so we have a separate table for each of these.

These ideas of entities and attributes are the same ideas that were discussed for XML design back in Section 5.5.2, just with different terminology.

Relationships

A **relationship** is an association between entities. For example, a publisher *publishes* books and a book *is published by* a publisher. Relationships are represented in a database by foreign key–primary key pairs, but the details depend on the **cardinality** of the relationship—whether the relationship is one-to-one, many-to-one, or many-to-many.

For example, a book is published by exactly one publisher, but a publisher publishes many books, so the relationship between books and publishers is many-to-one.

This sort of relationship can be represented by placing a foreign key in the table for books (the "many" side) that refers to the primary key in the table for publishers (the "one" side). This is the design that we have already seen, on page 122, where the `book_table` has a foreign key, `pub`, that refers to the primary key, `ID`, in the `publisher_table`.

One-to-one relationships can be handled similarly to many-to-one relationships (it does not matter which table gets the foreign key), but many-to-many relationships are more complex.

In our book database example, we can identify another sort of entity: authors.

In order to accommodate information about authors in the database, there should be another table for author information. In the example below, the table only contains the author's name, but other information, such as the author's age and nationality, could be added.

```
author_table  ( ID [PK], name )
```

What is the relationship between books and authors? An author can write several books and a book can have more than one author, so this is an example of a many-to-many relationship.

A many-to-many relationship can only be represented by creating a new table in the database.

For example, we can create a table, called the `book_author_table`, that contains the relationship between authors and books. This table contains a foreign key that refers to the author table and a foreign key that refers to the book table. The representation of book entities, author entities, and the relationship between them now consists of three tables, as shown below.

```
author_table  ( ID [PK], name )

book_table  ( ISBN [PK], title,
              pub [FK publisher_table.ID] )

book_author_table ( ID [PK],
                    book [FK book_table.ISBN],
                    author [FK author_table.ID] )
```

The book database design, with author information included, is shown in the diagram below.

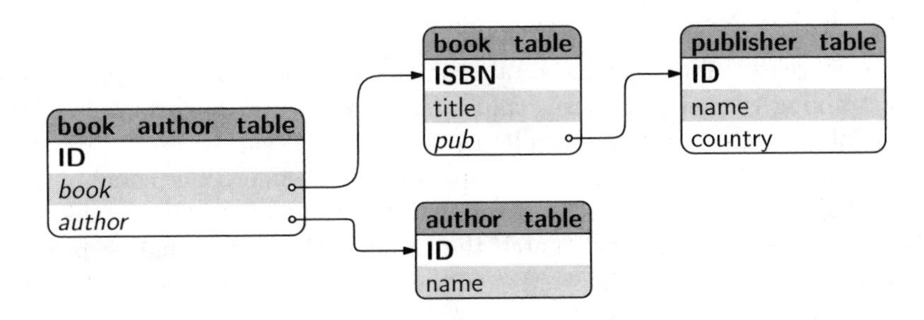

The contents of these tables for several books are shown below. The author table just lists the authors for whom we have information:

```
ID  name
--  -----------
2   Lynley Dodd
5   Eve Sutton
```

The `book_table` just lists the books that are in the database:

```
ISBN        title                                 pub
----------  ------------------------------------  ---
0908606664  Slinky Malinki                          1
1908606206  Hairy Maclary from Donaldson's Dairy  1
0908606273  My Cat Likes to Hide in Boxes           1
```

The `book_author_table` contains the association between books and authors:

```
ID  book        author
--  ----------  ------
2   0908606664  2
3   1908606206  2
6   0908606273  2
7   0908606273  5
```

Notice that author 2 (Lynley Dodd) has written more than one book and book 0908606273 (My Cat Likes to Hide in Boxes) has more than one author.

Designing for data integrity

Another reason for creating a table in a database is for the purpose of constraining the set of possible values for an attribute. For example, if the table of authors records the gender of the author, it can be useful to have a separate table that contains the possible values of gender. The column in the author table then becomes a foreign key referring to the gender table and, because a foreign key must match the value of the corresponding primary key, we have a check on the validity of the gender values in the author table.

The redesigned author table and gender table are described below.

```
author_table  ( ID [PK], name,
                gender [FK gender_table.ID] )

gender_table ( ID [PK], gender )
```

The `gender_table` only contains the set of possible gender values, as shown below.

```
ID  gender
--  ------
1   male
2   female
```

The final book database design, consisting of five tables, is shown in the diagram below.

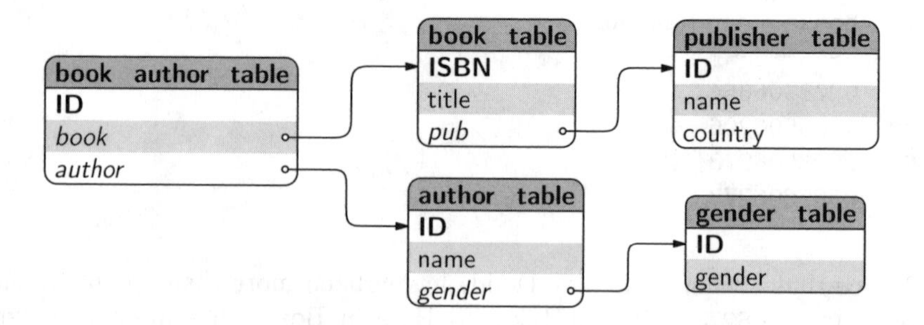

Database normalization

Another way to approach database design is to choose tables and columns within tables based on whether they satisfy a set of rules called **normal forms**.

This, more formal, process of database design is called **normalization**.

There are several normal forms, but we will only mention the first three because these will cover most simple situations.

The proper definition of normalization depends on more advanced relational database concepts that are beyond the scope of this book, so the descriptions below are just to give a feel for how the process works.

First normal form

First normal form requires that the columns in a table must be **atomic**, there should be no **duplicative columns**, and every table must have a primary key.

The first part of this rule says that a column in a database table must only contain a single value. As an example, consider the following possible design for a table for storing information about books. There is one column for the title of the book and another column for *all* authors of the book.

```
book_table ( title, authors )
```

Two rows of this table are shown below.

```
title                           authors
-----------------------------   -----------------------
Slinky Malinki                  Lynley Dodd
My Cat Likes to Hide in Boxes   Eve Sutton, Lynley Dodd
```

The first column of this table is acceptable because it just contains one piece of information: the title of the book. However, the second column is not atomic because it contains a *list* of authors for each book. The book on the second row has two authors recorded in the `authors` column. This violates first normal form.

The second part of the rule says that a table cannot have two columns containing the same information. For example, the following possible redesign of the book table provides a solution to the previous problem by having a separate column for each author of the book.

```
book_table ( title, author1, author2 )
```

Two rows from this table are shown below.

```
title                          author1     author2
------------------------------ ----------- -----------
Slinky Malinki                 Lynley Dodd NULL
My Cat Likes to Hide in Boxes  Eve Sutton  Lynley Dodd
```

This solves the problem of atomic columns because each column only contains the name of one author. However, the table has two duplicative columns: `author1` and `author2`. These two columns both record the same information, author names, so this design also violates first normal form.

A possible redesign that satisfies the requirement that each column is atomic and not duplicative is shown below. We now have just one column for the book title and one column for the names of the authors.

```
book_table ( title, author )
```

The contents of this table are shown below. Notice that the second book now occupies two rows because it has two authors.

```
title                          author
------------------------------ -----------
Slinky Malinki                 Lynley Dodd
My Cat Likes to Hide in Boxes  Eve Sutton
My Cat Likes to Hide in Boxes  Lynley Dodd
```

The final part of the first normal form rule says that there must be a column in the table that has a unique value in every row (or it must be possible to combine several columns to obtain a unique value for every row). In other words, every table must have a primary key.

Can we find a primary key in the table above?

Neither the `title` column nor the `author` column by itself is any use as a primary key because some values repeat in each of these columns.

We could combine the two columns to create a composite primary key. However, it is also important to think about not just the data that are currently in a table, but also what possible values could be entered into the table in the future (or even just in theory). In this case, it is possible that a book could be published in both hard cover and paperback formats, both of which would have the same title and author, so while a composite primary key would work for the three rows shown below, it is not necessarily a smart choice.

As described previously, for the case of information about books, a great candidate for a primary key is the book's ISBN because it is guaranteed to be unique for a particular book. If we add an `ISBN` column to the table, we can finally satisfy first normal form, though it still has to be a composite primary key involving the combination of `ISBN` and `author`.

```
book_table ( ISBN [PK],
             title,
             author [PK] )
```

The contents of this table are shown below.

```
ISBN        title                          author
----------  -----------------------------  -----------
0908606664  Slinky Malinki                 Lynley Dodd
0908606273  My Cat Likes to Hide in Boxes  Lynley Dodd
0908606273  My Cat Likes to Hide in Boxes  Eve Sutton
```

This is not an ideal solution for storing this information, but at least it satisfies first normal form. Consideration of second and third normal form will help to improve the design.

Second normal form

Second normal form requires that a table must be in first normal form *and* all columns in the table must relate to the *entire* primary key.

This rule formalizes the idea that there should be a table for each entity in the data set.

As a very basic example, consider the following table that contains information about authors and publishers. The primary key of this table is the author `ID`. In other words, each row of this table only concerns a single author.

```
author_table ( ID [PK], name, publisher)
```

Two rows from this table are shown below.

```
ID  name         publisher
--  -----------  ----------------
2   Lynley Dodd  Mallinson Rendel
5   Eve Sutton   Mallinson Rendel
```

The `name` column of this table relates to the primary key (the `ID`); this is the name of the author. However, the `publisher` column does not relate to the primary key. This is the publisher of a book. In other words, the information about publishers belongs in a table about publishers (or possibly a table about books), *not* in a table about authors.

As a more subtle example, consider the table that we ended up with at the end of first normal form.

```
book_table ( ISBN [PK],
             title,
             author [PK] )

ISBN        title                          author
----------  -----------------------------  -----------
0908606664  Slinky Malinki                 Lynley Dodd
0908606273  My Cat Likes to Hide in Boxes  Lynley Dodd
0908606273  My Cat Likes to Hide in Boxes  Eve Sutton
```

The primary key for this table is a combination of `ISBN` and `author` (each row of the table carries information about one author of one book).

The `title` column relates to the `ISBN`; this is the title of the book. However, the `title` column does not relate to the `author`; this is not the title of the author!

The table needs to be split into two tables, one with the information about books and one with the information about authors. Shown below is the book-related information separated into its own table.

```
book_table ( ISBN [PK],
             title )

ISBN        title
----------  -----------------------------
0908606664  Slinky Malinki
0908606273  My Cat Likes to Hide in Boxes
```

It is important to remember that each of the new tables that we create to satisfy second normal form must also satisfy first normal form. In this case, it would be wise to add an `ID` column to act as the primary key for the table of authors, as shown below, because it is entirely possible that two distinct authors could share the same name.

```
author_table ( ID [PK],
               author )
```

```
ID  author
--  -----------
2   Lynley Dodd
5   Eve Sutton
```

As this example makes clear, having split a table into two or more pieces, it is very important to link the pieces together by adding one or more foreign keys, based on the relationships between the tables. In this case, the relationship is many-to-many, so the solution requires a third table to provide a link between books and authors.

```
book_author_table ( ID [PK],
                    book [FK book_table.ISBN],
                    author [FK author_table.ID] )
```

```
ID  book        author
--  ----------  ------
2   0908606664  2
6   0908606273  2
7   0908606273  5
```

Third normal form

Third normal form requires that a table must be in second normal form *and* all columns in the table must relate *only* to the primary key (not to each other).

This rule further emphasizes the idea that there should be a separate table for each entity in the data set. For example, consider the following table for storing information about books.

```
book_table ( ISBN [PK],
             title,
             publisher,
             country )
```

```
ISBN        title             publisher         country
----------  ----------------  ----------------  -------
0395193958  The Hobbit        Houghton Mifflin  USA
0836827848  Slinky Malinki    Mallinson Rendel  NZ
0908783116  Mechanical Harry  Mallinson Rendel  NZ
```

The primary key of this table is the ISBN, which uniquely identifies a book. The `title` column relates to the book; this is the title of the book. Each row of this table is about one book.

The `publisher` column also relates to the book; this is the publisher of the book. However, the `country` column does not relate *directly* to the book; this is the country of the *publisher*. That obviously is information about the book—it is the country of the publisher of the book—but the relationship is indirect, through the publisher.

There is a simple heuristic that makes it easy to spot this sort of problem in a database table. Notice that the information in the `publisher` and `country` columns is identical for the books published by Mallinson Rendel. When two or more columns repeat the same information over and over, it is a sure sign that either second or third normal form is not being met.

In this case, the analysis of the table suggests that there should be a separate table for information about the publisher.

Applying the rules of normalization usually results in the creation of multiple tables in a database. The previous discussion of *relationships* should be consulted for making sure that any new tables are linked to at least one other table in the database using a foreign-key, primary-key pair.

Denormalization

The result of normalization is a well-organized database that should be easy to maintain. However, normalization may produce a database that is slow in terms of accessing the data (because the data from many tables has to be recombined).

Denormalization is the process of deliberately violating normal forms, typically in order to produce a database that can be accessed more rapidly.

5.6.4 Flashback: The DRY principle

A well-designed database, particularly one that satisfies third normal form, will have the feature that each piece of information is stored only once. Less repetition of data values means that a well-designed database will usually require less memory than storing an entire data set in a naïve single-table format. Less repetition also means that a well-designed database is easier to maintain and update, because if a change needs to be made, it only needs to be made in one location. Furthermore, there is less chance of errors creeping into the data set. If there are multiple copies of information, then it is possible for the copies to disagree, but with only one copy there can be no disagreements.

```
            VARIABLE : Mean TS from clear sky composite (kelvin)
            FILENAME : ISCCPMonthly_avg.nc
            FILEPATH : /usr/local/fer_dsets/data/
            SUBSET   : 24 by 24 points (LONGITUDE-LATITUDE)
            TIME     : 16-JAN-1995 00:00
             113.8W 111.2W 108.8W 106.2W 103.8W 101.2W 98.8W  ...
              27     28     29     30     31     32     33    ...
 36.2N / 51:  272.7  270.9  270.9  269.7  273.2  275.6  277.3  ...
 33.8N / 50:  279.5  279.5  275.0  275.6  277.3  279.5  281.6  ...
 31.2N / 49:  284.7  284.7  281.6  281.6  280.5  282.2  284.7  ...
 28.8N / 48:  289.3  286.8  286.8  283.7  284.2  286.8  287.8  ...
 26.2N / 47:  292.2  293.2  287.8  287.8  285.8  288.8  291.7  ...
 23.8N / 46:  294.1  295.0  296.5  286.8  286.8  285.2  289.8  ...
 ...
```

Figure 5.19: One of the plain text files from the original format of the Data Expo data set, which contains data for one variable for one month. The file contains information on latitude and longitude that is repeated in every other plain text file in the original format (for each variable and for each month; in total, over 500 times).

These ideas are an expression of the DRY principle from Section 2.7. A well-designed database is the ultimate embodiment of the DRY principle for data storage.

5.6.5 Case study: The Data Expo (continued)

The Data Expo data set consists of seven atmospheric variables recorded at 576 locations for 72 time points (every month for 6 years), plus elevation data for each location (see Section 5.2.8).

The data were originally stored as 505 plain text files, where each file contains the data for one variable for one month. Figure 5.19 shows the first few lines from one of the plain text files.

As we have discussed earlier in this chapter, this simple format makes the data very accessible. However, this is an example where a plain text format is quite inefficient, because many values are repeated. For example, the longitude and latitude information for each location in the data set is stored in every single file, which means that that information is repeated over 500 times! That not only takes up more storage space than is necessary, but it also violates the DRY principle, with all of the negative consequences that follow from that.

In this section, we will consider how the Data Expo data set could be stored as a relational database.

To start with, we will consider the problem from an entities and attributes perspective. What entities are there in the data set? In this case, the different entities that are being measured are relatively easy to identify. There are measurements on the *atmosphere*, and the measurements are taken at different *locations* and at different *times*. We have information about each time point (i.e., a date), we have information about each location (longitude and latitude and elevation), and we have several measurements on the atmosphere. This suggests that we should have three tables: one for atmospheric measures, one for locations, and one for time points.

It is also useful to look at the data set from a normalization perspective. For this purpose, we will start with all of the information in a single table (only 7 rows shown):

date	lon	lat	elv	chi	cmid	clo	ozone	press	stemp	temp
1995-01-16	-56.25	36.25	0.0	25.5	17.5	38.5	298.0	1000.0	289.8	288.8
1995-01-16	-56.25	33.75	0.0	23.5	17.5	36.5	290.0	1000.0	290.7	289.8
1995-01-16	-56.25	31.25	0.0	20.5	17.0	36.5	286.0	1000.0	291.7	290.7
1995-01-16	-56.25	28.75	0.0	12.5	17.5	37.5	280.0	1000.0	293.6	292.2
1995-01-16	-56.25	26.25	0.0	10.0	14.0	35.0	272.0	1000.0	296.0	294.1
1995-01-16	-56.25	23.75	0.0	12.5	11.0	32.0	270.0	1000.0	297.4	295.0
1995-01-16	-56.25	21.25	0.0	7.0	10.0	31.0	260.0	1000.0	297.8	296.5

In terms of first normal form, all columns are atomic and there are no duplicative columns, and we can, with a little effort, find a (composite) primary key: we need a combination of `date`, `lon` (longitude), and `lat` (latitude) to get a unique value for all rows.

Moving on to second normal form, the column `elv` (elevation) immediately fails. The elevation at a particular location clearly relates to the longitude and latitude of the location, but it has very little to do with the date. We need a new table to hold the longitude, latitude, and elevation data.

The new table design and the first three rows of data are shown below.

```
location_table ( longitude [PK],
                 latitude [PK],
                 elevation )
```

lon	lat	elv
-56.25	36.25	0.0
-56.25	33.75	0.0
-56.25	31.25	0.0

This "location" table is in third normal form. It has a primary key (a combination of longitude and latitude), and the elv column relates directly to the entire primary key.

Going back to the original table, the remaining columns of atmospheric measurements are all related to the primary key; the data in these columns represent an observation at a particular location at a particular time point.

However, we now have two tables rather than just one, so we must make sure that the tables are linked to each other, and in order to achieve this, we need to determine the relationships between the tables.

We have two tables, one representing atmospheric measurements, at various locations and times, and one representing information about the locations. What is the relationship between these tables? Each location (each row of the location table) corresponds to several measurements, but each individual measurement (each row of the measurement table) corresponds to only one location, so the relationship is many-to-one.

This means that the table of measurements should have a foreign key that references the primary key in the location table. The design could be expressed like this:

```
location_table ( longitude [PK],
                 latitude [PK],
                 elevation )

measure_table ( date [PK],
                longitude [PK] [FK location_table.longitude],
                latitude [PK] [FK location_table.latitude],
                cloudhigh, cloudlow, cloudmid, ozone,
                pressure, surftemp, temperature )
```

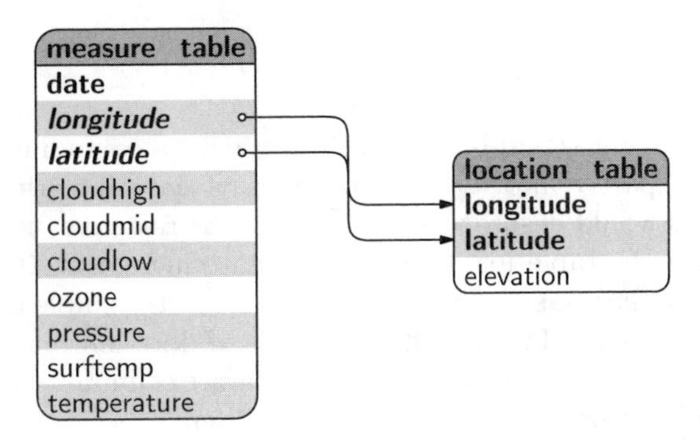

Both tables have composite primary keys. The measure_table also has a composite foreign key, to refer to the composite primary key in the location_table. Finally, the longitude and latitude columns have roles in both the primary key and the foreign key of the measure_table.

A possible adjustment to the database design is to consider a surrogate auto-increment key—a column that just corresponds to the row number in the table—as the primary key for the location table, because the natural primary key is quite large and cumbersome. This leads to a final design that can be expressed as below.

```
location_table ( ID [PK],
                 longitude, latitude, elevation )

measure_table ( date [PK],
                location [PK] [FK location_table.ID],
                cloudhigh, cloudlow, cloudmid, ozone,
                pressure, surftemp, temperature )
```

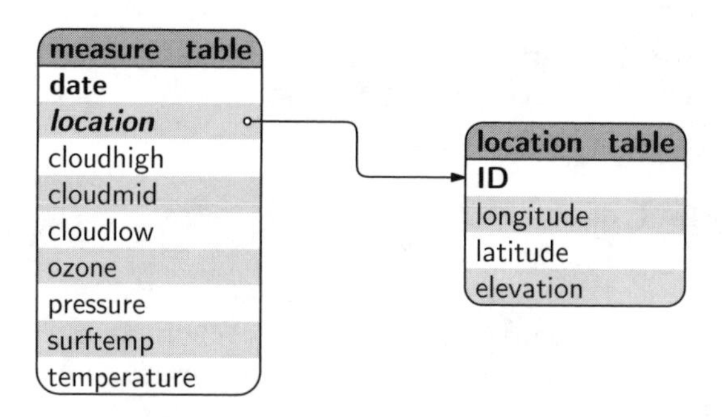

Another adjustment would be to break out the date column into a separate table. This is partly motivated by the idea of data integrity; a separate table for dates would ensure that all dates in the measure_table are valid dates. Also, if the table for dates uses an auto-increment ID column, the date column in the measure_table can become just a simple integer, rather than a lengthy date value. Finally, the table of date information can have the year and month information split into separate columns, which can make it more useful to work with the date information.

The final Data Expo database design is shown below.

```
date_table ( ID [PK], date, month, year )

location_table ( ID [PK],
                 longitude, latitude, elevation )

measure_table ( date [PK] [FK date_table.ID],
                location [PK] [FK location_table.ID],
                cloudhigh, cloudlow, cloudmid, ozone,
                pressure, surftemp, temperature )
```

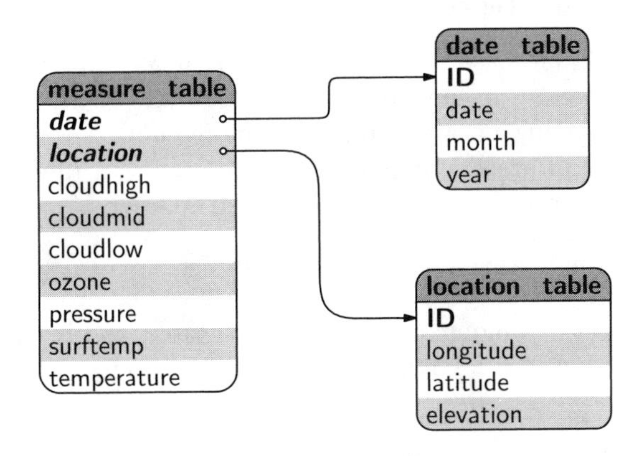

As a final check, we should confirm that these tables all satisfy third normal form.

Each table has a primary key, all columns are atomic, and there are no duplicative columns, so first normal form is satisfied. All of the columns in each table correspond to the primary key of the table—in particular, each measurement in the `measure_table` corresponds to a particular combination of date and location—so second normal form is also satisfied. The tables mostly also satisfy third normal form because columns generally relate *only* to the primary key in the table. However, it could be argued that, in the `date_table`, the `month` and `year` columns relate to the `date` column as well as to the primary key of the table. This is a good demonstration of a possible justification for denormalization; we have split out these columns because we anticipate that they will be useful for asking questions of the database in the future. The ideas of normalization should be used as guides for achieving a sensible database design, but other considerations may also come into play.

5.6.6 Advantages and disadvantages

The previous sections have demonstrated that databases are a lot more complex than most of the other data storage options in this chapter. In this section, we will look at what we can gain by using a database to store a data set and what the costs are compared to other storage formats.

The relatively formal data model of relational databases, and the relatively complex processes that go into designing an appropriate database structure, are worthwhile because the resulting structure enforces constraints on the data in a database, which means that there are checks on the accuracy and consistency of data that are stored in a database. In other words, databases ensure better **data integrity**.

For example, the database structure ensures that all values in a single column of a table are of the same data type (e.g., they are all numbers). It is possible, when setting up a database, to enforce quite specific constraints on what values can appear in a particular column of a table. Section 8.3 provides some information on this topic of the creation of data sets.

Another important structural feature of databases is the existence of foreign keys and primary keys. Database software will enforce the rule that a primary key must be unique for every row in a table, and it will enforce the rule that the value of a foreign key must refer to an existing primary key value.

Databases tend to be used for large data sets because, for most DBMS, there is no limit on the size of a database. However, even when a data set is not enormous, there are advantages to using a database because the organization of the data can improve accuracy and efficiency. In particular, databases allow the data to be organized in a variety of ways so that, for example, data with a hierarchical structure can be stored in an efficient and natural way.

Databases are also advantageous because most DBMS provide advanced features that are far beyond what is provided by the software that is used to work with data in other formats (e.g., text editors and spreadsheet programs). These features include the ability to allow multiple people to access and even modify the data at once and advanced security to control who has access to the data and who is able to modify the data.

The first *cost* to consider is monetary. The commercial database systems offered by Oracle and Microsoft can be very expensive, although open source options exist (see Section 5.6.9) to relieve that particular burden. However, there is also the cost of acquiring or hiring the expertise necessary to create, maintain, and interact with data stored in a database.

Another disadvantage of using a database as a storage format is that the data can only be accessed using a specific piece of DBMS software.

Finally, all of the sophistication and flexibility that a database provides may just not be necessary for small data sets or for data sets that have a simple structure. For example, a binary format such as netCDF is very well suited to a geographical data set where observations are made on a regular grid of locations and at a fixed set of time points and it will outperform a more general-purpose database solution.

The investment required to create and maintain a database means that it will not always be an appropriate choice.

5.6.7 Flashback: Database design and XML design

In Section 5.5.2 we discussed some basic ideas for deciding how to represent a data set in an XML format.

The ideas of database design that we have discussed in Section 5.6.3—entities, attributes, relationships, and normalization—are very similar to the ideas from XML design, if a little more formal.

This similarity arises from the fact that we are trying to solve essentially the same problem in both cases, and this can be reflected in a simple correspondence between database designs and XML designs for the same data set.

As a rough guideline, a database table can correspond to a set of XML elements of the same type. Each row of the table will correspond to a single XML element, with each column of values recorded as a separate attribute within the element. The caveats about when attributes cannot be used still apply (see page 108).

Simple one-to-one or many-to-one relationships can be represented in XML by nesting several elements (the many) within another element (the one). More complex relationships cannot be solved by nesting, but attributes corresponding to primary keys and foreign keys can be used to emulate relationships between entities via XML elements that are not nested.

5.6.8 Case study: The Data Expo (continued)

The Data Expo data set consists of several atmospheric measurements taken at many different locations and at several time points. A database design that we developed for storing these data consisted of three tables: one for the location data, one for the time data, and one for the atmospheric measurements (see Section 5.6.5). The database schema is reproduced below for easy reference.

```
date_table ( ID [PK], date, month, year )

location_table ( ID [PK],
                 longitude, latitude, elevation )

measure_table ( date [PK] [FK date_table.ID],
                location [PK] [FK location_table.ID],
                cloudhigh, cloudlow, cloudmid, ozone,
                pressure, surftemp, temperature )
```

We can translate this database design into an XML document design very simply, by creating a set of elements for each table, with attributes for each column of data. For example, the fact that there is a table for location information implies that we should have `location` elements, with an attribute for each column in the database table. The data for the first few locations are represented like this in a database table:

```
ID      lon     lat  elv
------  -----   ---  ----------
1       -113.   36.  1526.25
2       -111.   36.  1759.56
3       -108.   36.  1948.38
```

The same data could be represented in XML like this:

```
<location id="1" longitude="-113.75" latitude="36.25"
          elevation="1526.25" />
<location id="2" longitude="-111.25" latitude="36.25"
          elevation="1759.56" />
<location id="3" longitude="-108.75" latitude="36.25"
          elevation="1948.38" />
```

As an analogue of the primary keys in the database design, the DTD for this XML design could specify `id` as an `ID` attribute (see Section 6.2.2).

An XML element for the first row from the `date_table` might look like this (again with `id` as an `ID` attribute in the DTD):

```
<date id="1" date="1995-01-16"
      month="January" year="1995" />
```

Because there is a many-to-many relationship between locations and dates, it would not make sense to nest the corresponding XML elements. Instead, the XML elements that correspond to the rows of the `measure_table` could include attributes that refer to the relevant location and date elements. The following code shows an example of what a `measure` XML element might look like.

```
<measure date="1" location="1"
         cloudhigh="26.0" cloudmid="34.5"
         cloudlow="7.5" ozone="304.0"
         pressure="835.0" surftemp="272.7"
         temperature="272.1" />
```

In order to enforce the data integrity of the attributes `date` and `location`, the DTD for this XML design would specify these as `IDREF` attributes (see Section 6.2.2).

5.6.9 Database software

Every different database software product has its own format for storing the database tables on disk, which means that data stored in a database are only accessible via one specific piece of software.

This means that, if we are given data stored in a particular database format, we are forced to use the corresponding software. Something that slightly alleviates this problem is the existence of a standard language for querying databases. We will meet this language, SQL, in the next chapter.

If we are in the position of storing information in a database ourselves, there are a number of fully featured open source database management systems to choose from. PostgreSQL[6] and MySQL[7] are very popular options, though they require some investment in resources and expertise to set up because they have separate client and server software components. SQLite[8] is much simpler to set up and use, especially for a database that only requires access by a single person working on a single computer.

Section 7.2.14 provides a very brief introduction to SQLite.

The major proprietary database systems include Oracle, Microsoft SQL Server, and Microsoft Access. The default user interface for these software products is based on menus and dialogs so they are beyond the scope and

[6]`http://www.postgresql.org/`
[7]`http://www.mysql.com/`
[8]`http://www.sqlite.org/`

interest of this book. Nevertheless, in all of these, as with the default interfaces for the open source database software, it is possible to write computer code to access the data. Writing these **data queries** is the topic of the next chapter.

Recap

A database consists of one or more tables. Each column of a database table contains only one type of information, corresponding to one variable from a data set.

A primary key uniquely identifies each row of a table. A primary key is a column in a table with a different value on every row.

A foreign key relates one table to another within a database. A foreign key is a column in a table that refers to the values in the primary key of another table.

A database should be designed so that information about different entities resides in separate tables.

Normalization is a way to produce a good database design.

Databases can handle large data sets and data sets with a complex structure, but databases require specific software and a certain level of expertise.

5.7 Further reading

Modern Database Management
by Jeffrey A. Hoffer, Mary Prescott, and Fred McFadden
7th edition (2004) Prentice Hall.
Comprehensive textbook treatment of databases and associated technologies, with more of a business focus. Includes many advanced topics beyond the scope of this book.

Summary

Simple text data is stored using 1 byte per character. Integers are stored using 2 or 4 bytes and real values typically use 4 or 8 bytes.

There is a limit to the size of numbers that can be stored digitally and for real values there is a limit on the precision with which values can be stored.

Plain text files are the simplest data storage solution, with the advantage that they are simple to use, work across different computer platforms, and work with virtually any software. The main disadvantage to plain text files is their lack of standard structure, which means that software requires human input to determine where data values reside within the file. Plain text files are also generally larger and slower than other data storage options.

CSV (comma-separated values) files offer the most standardized plain text format.

Binary formats tend to provide smaller files and faster access speeds. The disadvantage is that data stored in a binary format can only be accessed using specific software.

Spreadsheets are ubiquitous, flexible, and easy to use. However, they lack structure so should be used with caution.

XML is a language that can be used for marking up data. XML files are plain text but provide structure that allows software to automatically determine the location of data values within the file (XML files are self-describing).

Databases are sophisticated but relatively complex. They are useful for storing very large or very complex data sets but require specific software and much greater expertise.

6

XML Reference

XML (the eXtensible Markup Language) is a data description language that can be used for storing data. It is particularly useful as a format for sharing information between different software systems.

The information in this chapter describes XML 1.0, which is a W3C Recommendation.

Within this chapter, any code written in a *sans-serif oblique font* represents a general template; that part of the code will vary depending on the names of the elements and the names of the attributes that are used to store a particular data set.

XML

6.1 XML syntax

The first line of an XML document should be a declaration that this is an XML document, including the version of XML being used.

```
<?xml version="1.0"?>
```

It is also useful to include a statement of the encoding used in the file.

```
<?xml version="1.0" encoding="UTF-8"?>
```

The main content of an XML document consists entirely of XML **elements**. An element usually consists of a start **tag** and an end tag, with plain text content or other XML elements in between.

A start tag is of the form <*elementName*> and an end tag has the form </*elementName*>.

The following code shows an example of an XML element.

```
<filename>ISCCPMonthly_avg.nc</filename>
```

The components of this XML element are shown below.

element:	<filename>ISCCPMonthly_avg.nc</filename>
start tag:	<filename>ISCCPMonthly_avg.nc</filename>
content:	<filename>ISCCPMonthly_avg.nc</filename>
end tag:	<filename>ISCCPMonthly_avg.nc</filename>

The start tag may include **attributes** of the form *attrName*="*attrValue*". The attribute value must be enclosed within double-quotes.

The names of XML elements and XML attributes are case-sensitive.

It is also possible to have an empty element, which consists of a single tag, with attributes. In this case, the tag has the form <*elementName* />.

The following code shows an example of an empty XML element with two attributes.

```
<case date="16-JAN-1994"
      temperature="278.9" />
```

The components of this XML element are shown below.

element name:	<case date="16-JAN-1994"
attribute name:	<case date="16-JAN-1994"
attribute value:	<case date="16-JAN-1994"
attribute name:	temperature="278.9" />
attribute value:	temperature="278.9" />

XML elements may be nested; an XML element may have other XML elements as its content. An XML document must have a single **root element**, which contains all other XML elements in the document.

The following code shows a very small, but complete, XML document. The root element of this document is the `temperatures` element. The `filename` and `case` elements are nested within the `temperatures` element.

```
<?xml version="1.0"?>
<temperatures>
    <filename>ISCCPMonthly_avg.nc</filename>
    <case date="16-JAN-1994"
          temperature="278.9"/>
</temperatures>
```

Table 6.1: The predefined XML entities.

Character	Description	Entity
<	less-than sign	<
>	greater-than sign	>
&	ampersand	&
"	quotation mark	"
'	apostrophe	'

A comment in XML is anything between the delimiters <!-- and -->.

For the benefit of human readers, the contents of an XML element are usually indented. However, whitespace is preserved within XML so this is not always possible when including plain text content.

In XML code, certain characters, such as the greater-than and less-than signs, have special meanings. Table 6.1 lists these special characters and also gives the escape sequence required to produce the normal, literal meaning of the characters.

A special syntax is provided for escaping an entire section of plain text content for the case where many such special characters are included. Any text between the delimiters <![CDATA[and]]> is treated as literal.

6.2 Document Type Definitions

An XML document that obeys the rules of the previous section is described as **well-formed**.

It is also possible to specify additional rules for the structure and content of an XML document, via a **schema** for the document. If the document is well-formed and also obeys the rules given in a schema, then the document is described as **valid**.

The Document Type Definition language (DTD) is a language for describing the schema for an XML document. DTD code consists of **element declarations** and **attribute declarations**.

XML

6.2.1 Element declarations

An element declaration should be included for every different type of element that will occur in an XML document. Each declaration describes what content is allowed inside a particular element. An element declaration is of the form:

```
<!ELEMENT elementName elementContents>
```

The *elementContents* specifies whether an element can contain plain text, or other elements (and if so, which elements, in what order), or whether the element must be empty. Some possible values are shown below.

`EMPTY`
: The element is empty.

`ANY`
: The element may contain anything (other elements, plain text, or both).

`(#PCDATA)`
: The element may contain plain text.

`(`*elementA*`)`
: The element must contain exactly one *elementA* element. The parentheses, `(` and `)`, are essential in this example and all others below.

`(`*elementA*`*)`
: The element may contain zero or more *elementA* elements. The asterisk, `*`, indicates "zero or more".

`(`*elementA*`+)`
: The element must contain one or more *elementA* elements. The plus sign, `+`, indicates "one or more".

`(`*elementA*`?)`
: The element must contain zero or one *elementA* elements. The question mark, `?`, indicates "zero or one".

`(`*elementA*`,`*elementB*`)`
: The element must contain exactly one *elementA* element and exactly one *elementB* element. The element names are separated from each other by commas.

`(`*elementA*`|`*elementB*`)`
: The element must contain either exactly one *elementA* element *or* exactly one *elementB* element. The vertical bar, `|`, indicates alternatives.

`(#PCDATA|elementA|elementB*)`
: The element may contain plain text, or a single *elementA* element, or zero or more *elementB* elements. The asterisk, `*`, is *inside* the parentheses so only applies to the *elementB* element.

`(#PCDATA|elementA|elementB)*`
: The element may contain plain text, plus zero or more occurrences of *elementA* elements and *elementB* elements. The asterisk, `*`, is *outside* the parentheses so applies to all elements within the parentheses.

6.2.2 Attribute declarations

An attribute declaration should be included for every different type of element that can have attributes. The declaration describes which attributes an element may have, what sorts of values the attribute may take, and whether the attribute is optional or compulsory. An attribute declaration is of the form:

```
<!ATTLIST elementName
    attrName attrType attrDefault
    ...
>
```

The *attrType* controls what value the attribute can have. It can have one of the following forms:

`CDATA`
: The attribute can take any value. Attribute values must always be plain text and escape sequences (XML entities) must be used for special XML characters (see Table 6.1).

`ID`
: The value of this attribute must be unique for all elements of this type in the document (i.e., a unique identifier). This is similar to a primary key in a database table.

  The value of an `ID` attribute must *not* start with a digit.

`IDREF`
: The value of this attribute must be the value of some other element's `ID` attribute. This is similar to a foreign key in a database table.

`(option1|option2)`
: This form provides a list of the possible values for the attribute. The

list of options is given, separated by vertical bars, `|`. This is a good way to limit an attribute to only valid values (e.g., only `"male"` or `"female"` for a `gender` attribute).

```
<!ATTLIST elementName
    gender (male|female) #REQUIRED>
```

The *attrDefault* either provides a default value for the attribute or states whether the attribute is optional or required (i.e., must be specified). It can have one of the following forms:

value
: This is the default value for the attribute.

`#IMPLIED`
: The attribute is optional. It is valid for elements of this type to contain this attribute, but it is not required.

`#REQUIRED`
: The attribute is required so it must appear in all elements of this type.

6.2.3 Including a DTD

A DTD can be embedded within an XML document or the DTD can be located within a separate file and referred to from the XML document.

The DTD information is included within a `DOCTYPE` declaration following the XML declaration. An inline DTD has the form:

```
<!DOCTYPE rootElementName [
    DTD code
]>
```

An external DTD stored in a file called `file.dtd` would be referred to as follows:

```
<!DOCTYPE rootElementName SYSTEM "file.dtd">
```

The name following the keyword `DOCTYPE` must match the name of the root element in the XML document.

```
 1 <?xml version="1.0"?>
 2 <!DOCTYPE temperatures [
 3     <!ELEMENT temperatures (filename, case)>
 4     <!ELEMENT filename (#PCDATA)>
 5     <!ELEMENT case EMPTY>
 6     <!ATTLIST case
 7         date        CDATA  #REQUIRED
 8         temperature CDATA  #IMPLIED>
 9 ]>
10 <temperatures>
11     <filename>ISCCPMonthly_avg.nc</filename>
12     <case date="16-JAN-1994"
13           temperature="278.9"/>
14 </temperatures>
```

Figure 6.1: A well-formed and valid XML document, with an embedded DTD. The line numbers (in grey) are just for reference.

6.2.4 An example

Figure 6.1 shows a very small, *well-formed* and *valid* XML document with an embedded DTD.

Line 1 is the required XML declaration.

Lines 2 to 9 provide a DTD for the document. This DTD specifies that the root element for the document must be a `temperatures` element (line 2). The `temperatures` element must contain one `filename` element and one `case` element (line 3). The `filename` element must contain only plain text (line 4) and the `case` element must be empty (line 5).

The `case` element must have a `date` attribute (line 7) and may also have a `temperature` attribute (line 8). The values of both attributes can be arbitrary text (`CDATA`).

The elements within the XML document that mark up the actual data values are on lines 10 to 14.

6.3 Further reading

The W3C XML 1.0 Specification
: `http://www.w3.org/TR/2006/REC-xml-20060816/`
The formal and official definition of XML. Quite technical.

The w3schools XML Tutorial
: `http://www.w3schools.com/xml/`
Quick, basic tutorial-based introduction to XML.

The w3schools DTD Tutorial
: `http://www.w3schools.com/dtd/`
Quick, basic tutorial-based introduction to DTDs.

The W3C Validation Service
: `http://validator.w3.org/`
This will check raw XML files as well as HTML documents.

libxml2
: `http://xmlsoft.org/`
This software includes a command-line tool, `xmllint` for checking XML code, including validating it against a DTD.

7
Data Queries

Having stored information in a particular data format, how do we get it back out again? How easy is it to access the data? The answer naturally depends on which data **format** we are dealing with.

For data stored in plain text files, it is very easy to find software that can read the files, although the software may have to be provided with additional information about the structure of the files—where the data values reside within the file—plus information about what sorts of values are stored in the file—whether the data are, for example, numbers or text.

For data stored in binary files, the main problem is finding software that is designed to read the specific binary format. Having done that, the software does all of the work of extracting the appropriate data values. This is an all or nothing scenario; we either have software to read the file, in which case data extraction is trivial, or we do not have the software, in which case we can do nothing. This scenario includes most data that are stored in spreadsheets, though in that case the likelihood of having appropriate software is much higher.

Another factor that determines the level of difficulty involved in retrieving data from storage is the *structure* of the data within the data format.

Data that are stored in plain text files, spreadsheets, or binary formats typically have a straightforward structure. For example, all of the values in a single variable within a data set are typically stored in a single column of a text file or spreadsheet, or within a single block of memory within a binary file.

By contrast, data that have been stored in an XML document or in a relational database can have a much more complex structure. Within XML files, the data from a single variable may be represented as attributes spread across several different elements, and data that are stored in a database may be spread across several tables.

This means that it is not necessarily straightforward to extract data from an XML document or a relational database. Fortunately, this is offset by the fact that sophisticated technologies exist to support data queries with relational databases and XML documents.

How this chapter is organized

To begin with, we will look at a simple example of data retrieval from a database. As with previous introductory examples, the focus at this point is not so much on the computer code itself as it is on the concepts involved and what sorts of tasks we are able to perform.

The main focus of this chapter is Section 7.2 on the **Structured Query Language** (**SQL**), the language for extracting information from relational databases. We will also touch briefly on **XPath** for extracting information from XML documents in Section 7.3.

7.1 Case study: The Data Expo (continued)

The Data Expo data set consists of seven atmospheric measurements at locations on a 24 by 24 grid averaged over each month for six years (72 time points). The elevation (height above sea level) at each location is also included in the data set (see Section 5.2.8 for more details).

The data set was originally provided as 505 plain text files, but the data can also be stored in a database with the following structure (see Section 5.6.5).

```
date_table ( ID [PK], date, month, year )

location_table ( ID [PK],
                 longitude, latitude, elevation )

measure_table ( date [PK] [FK date_table.ID],
                location [PK] [FK location_table.ID],
                cloudhigh, cloudlow, cloudmid, ozone,
                pressure, surftemp, temperature )
```

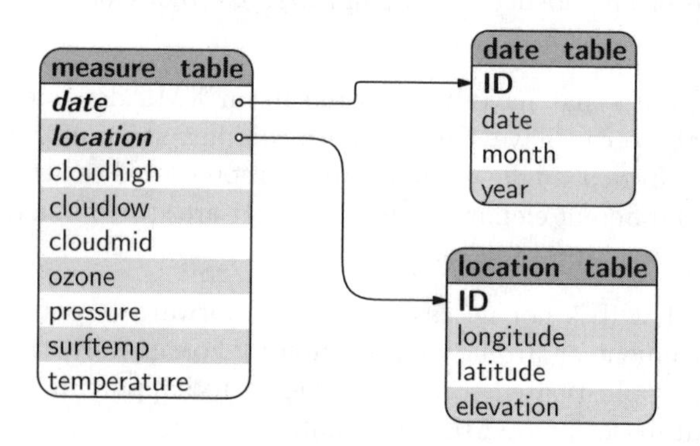

The `location_table` contains all of the geographic locations at which measurements were taken, and includes the elevation at each location.

```
ID  longitude  latitude  elevation
--  ---------  --------  ---------
1   -113.75    36.25     1526.25
2   -111.25    36.25     1759.56
3   -108.75    36.25     1948.38
4   -106.25    36.25     2241.31
...
```

The `date_table` contains all of the dates at which measurements were taken. This table also includes the text form of each month and the numeric form of the year. These have been split out to make it easier to perform queries based on months or years. The full dates are stored using the ISO 8601 format so that alphabetical ordering gives chronological order.

```
ID  date        month     year
--  ----------  --------  ----
1   1995-01-16  January   1995
2   1995-02-16  February  1995
3   1995-03-16  March     1995
4   1995-04-16  April     1995
...
```

The `measure_table` contains all of the atmospheric measurements for all dates and locations. Dates and locations are represented by simple ID numbers, referring to the appropriate complete information in the `date_table` and `location_table`. In the output below, the column names have been abbreviated to save space.

```
loc  date  chigh  cmid  clow  ozone  press  stemp  temp
---  ----  -----  ----  ----  -----  -----  -----  -----
1    1     26.0   34.5  7.5   304.0  835.0  272.7  272.1
2    1     23.0   32.0  7.0   306.0  810.0  270.9  270.3
3    1     23.0   32.0  7.0   306.0  810.0  270.9  270.3
4    1     17.0   29.5  7.0   294.0  775.0  269.7  270.9
...
```

With the data stored in this way, how difficult is it to extract information?

Some things are quite simple. For example, it is straightforward to extract all of the ozone measurements from the `measure_table`. The following SQL code performs this step.

```
SQL> SELECT ozone FROM measure_table;

ozone
-----
304.0
306.0
306.0
294.0
...
```

Throughout this chapter, examples of SQL code will be displayed like this, with the SQL code preceded by a **prompt**, `SQL>`, and the output from the code—the data that have been extracted from the database—displayed below the code, in a tabular format.

This information is more useful if we also know where and when each ozone measurement was taken. Extracting this additional information is also not difficult because we can just ask for the `location` and `date` columns as well.

```
SQL> SELECT date, location, ozone FROM measure_table;

date  location  ozone
----  --------  -----
1     1         304.0
1     2         306.0
1     3         306.0
1     4         294.0
...
```

Unfortunately, this is still not very useful because a `date` or `location` of `1` does not have a clear intuitive meaning. What we need to do is combine the values from the three tables in the database so that we can, for example, resolve the date value `1` to the corresponding real date `1995-01-16`.

This is where the extraction of information from a database gets interesting—when information must be combined from more than one table.

In the following code, we extract the `date` column from the `date_table`, the longitude and latitude from the `location_table`, and the `ozone` from the `measure_table`. Combining information from multiple tables like this is called a **database join**.

```
SQL> SELECT dt.date date,
            lt.longitude long, lt.latitude lat,
            ozone
         FROM measure_table mt
             INNER JOIN date_table dt
                 ON mt.date = dt.ID
             INNER JOIN location_table lt
                 ON mt.location = lt.ID;


date         long     lat   ozone
----------   -------  ----- -----
1995-01-16   -113.75  36.25 304.0
1995-01-16   -111.25  36.25 306.0
1995-01-16   -108.75  36.25 306.0
1995-01-16   -106.25  36.25 294.0
...
```

This complex code is one of the costs of having data stored in a database, but if we learn a little SQL so that we can do this sort of fundamental task, we gain the benefit of the wider range of capabilities that SQL provides. As a simple example, the above task can be modified very easily if we want to only extract ozone measurements from the first location (the difference is apparent in the result because the date values change, while the locations remain the same).

```
SQL> SELECT dt.date date,
            lt.longitude long, lt.latitude lat,
            ozone
         FROM measure_table mt
             INNER JOIN date_table dt
                 ON mt.date = dt.ID
             INNER JOIN location_table lt
                 ON mt.location = lt.ID
         WHERE mt.location = 1;


date         long     lat   ozone
----------   -------  ----- -----
1995-01-16   -113.75  36.25 304.0
1995-02-16   -113.75  36.25 296.0
1995-03-16   -113.75  36.25 312.0
1995-04-16   -113.75  36.25 326.0
...
```

In this chapter we will gain these two useful skills: how to use SQL to perform necessary tasks with a database—the sorts of things that are quite straightforward with other storage formats—*and* how to use SQL to perform tasks with a database that are much more sophisticated than what is possible with other storage options.

7.2 Querying databases

SQL is a language for creating, configuring, and querying relational databases. It is an open standard that is implemented by all major DBMS software, which means that it provides a consistent way to communicate with a database no matter which DBMS software is used to store or access the data.

Like all languages, there are different versions of SQL. The information in this chapter is consistent with SQL-92.

SQL consists of three components:

Data Definition Language (DDL)
: This is concerned with the creation of databases and the specification of the structure of tables and of constraints between tables. This part of the language is used to specify the data types of each column in each table, which column(s) make up the primary key for each table, and how foreign keys are related to primary keys. We will not discuss this part of the language in this chapter, but some mention of it is made in Section 8.3.

Data Control Language (DCL)
: This is concerned with controlling access to the database—*who* is allowed to do *what* to *which* tables. This part of the language is the domain of database administrators and need not concern us.

Data Manipulation Language (DML)
: This is concerned with getting data into and out of a database and is the focus of this chapter.

In this section, we are only concerned with one particular command within the DML part of SQL: the `SELECT` command for extracting values from tables within a database.

Section 8.3 includes brief information about some of the other features of SQL.

7.2.1 SQL syntax

Everything we do in this section will be a variation on the `SELECT` command of SQL, which has the following basic form:

```
SELECT columns
    FROM tables
    WHERE row_condition
```

This will extract the specified *columns* from the specified *tables* but it will only include the rows for which the *row_condition* is true.

The keywords `SELECT`, `FROM`, and `WHERE` are written in uppercase by convention and the names of the columns and tables will depend on the database that is being queried.

Throughout this section, SQL code examples will be presented after a "prompt", `SQL>`, and the result of the SQL code will be displayed below the code.

7.2.2 Case study: The Data Expo (continued)

The goal for contestants in the Data Expo was to summarize the important features of the atmospheric measurements. In this section, we will perform some straightforward explorations of the data in order to demonstrate a variety of simple SQL commands.

A basic first step in data exploration is just to view the univariate distribution of each measurement variable. The following code extracts all air pressure values from the database using a very simple SQL query that selects all rows from the `pressure` column of the `measure_table`.

```
SQL> SELECT pressure FROM measure_table;
```

```
pressure
--------
835.0
810.0
810.0
775.0
795.0
915.0
...
```

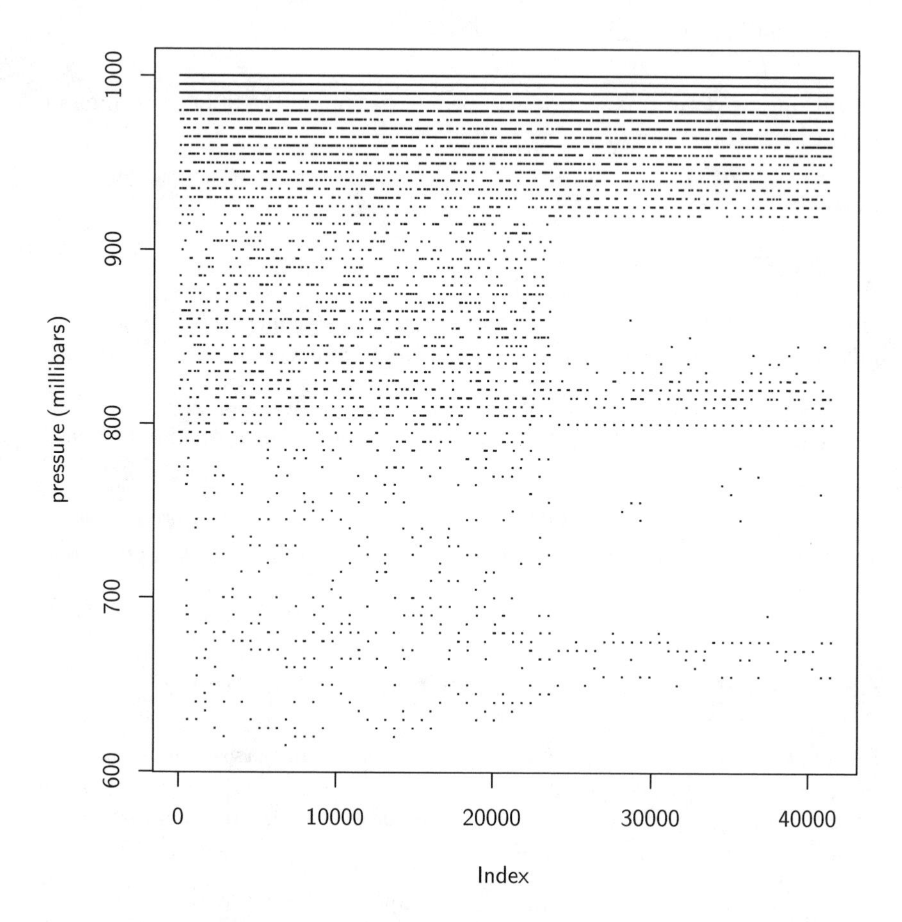

Figure 7.1: A plot of the air pressure measurements from the 2006 JSM Data Expo. This includes pressure measurements at all locations and at all time points.

This SELECT command has no restriction on the rows, so the result contains *all* rows from the table. There are 41,472 ($24 \times 24 \times 72$) rows in total, so only the first few are shown here. Figure 7.1 shows a plot of all of the pressure values.

The resolution of the data is immediately apparent; the pressure is only recorded to the nearest multiple of 5. However, the more striking feature is the change in the spread of the second half of the data. NASA has confirmed that this change is real but unfortunately has not been able to give an explanation for why it occurred.

An entire column of data from the measure_table in the Data Expo database represents measurements of a single variable at *all* locations for *all* time periods. One interesting way to "slice" the Data Expo data is to look at the values for a *single* location over all time periods. For example, how does surface temperature vary over time at a particular location?

The following code shows a slight modification of the previous query to obtain a different column of values, `surftemp`, and to only return some of the rows from this column. The `WHERE` clause limits the result to rows for which the `location` column has the value `1`.

```
SQL> SELECT surftemp
         FROM measure_table
         WHERE location = 1;


surftemp
--------
272.7
282.2
282.2
289.8
293.2
301.4
...
```

Again, the result is too large to show all values, so only the first few are shown. Figure 7.2 shows a plot of all of the values.

The interesting feature here is that we can see a cyclic change in temperature, as we might expect, with the change of seasons.

The order of the rows in a database table is not guaranteed. This means that whenever we extract information from a table, we should be explicit about the order in which we want for the results. This is achieved by specifying an `ORDER BY` clause in the query. For example, the following SQL command extends the previous one to ensure that the temperatures for location `1` are returned in chronological order.

```
SELECT surftemp
    FROM measure_table
    WHERE location = 1
    ORDER BY date;
```

The `WHERE` clause can use other comparison operators besides equality. As a trivial example, the following code has the same result as the previous

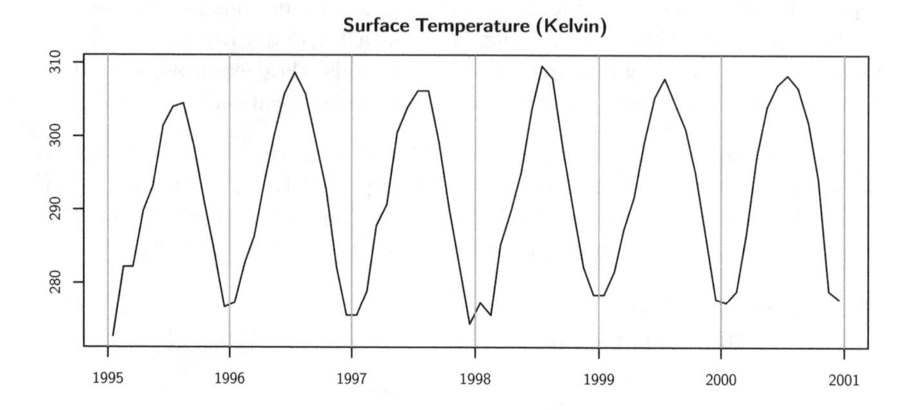

Figure 7.2: A plot of the surface temperature measurements from the 2006 JSM Data Expo, for all time points, at location `1`. Vertical grey bars mark the change of years.

example by specifying that we only want rows where the `location` is less than 2 (the only location value less than two is the value `1`).

```
SELECT surftemp
    FROM measure_table
    WHERE location < 2
    ORDER BY date;
```

It is also possible to combine several conditions within the `WHERE` clause, using logical operators `AND`, to specify conditions that must *both* be true, and `OR`, to specify that we want rows where *either* of two conditions are true. As an example, the following code extracts the surface temperature for two locations. In this example, we include the `location` and `date` columns in the result to show that rows from both locations (for the same date) are being included in the result.

```
SQL> SELECT location, date, surftemp
         FROM measure_table
         WHERE location = 1 OR
               location = 2
         ORDER BY date;
```

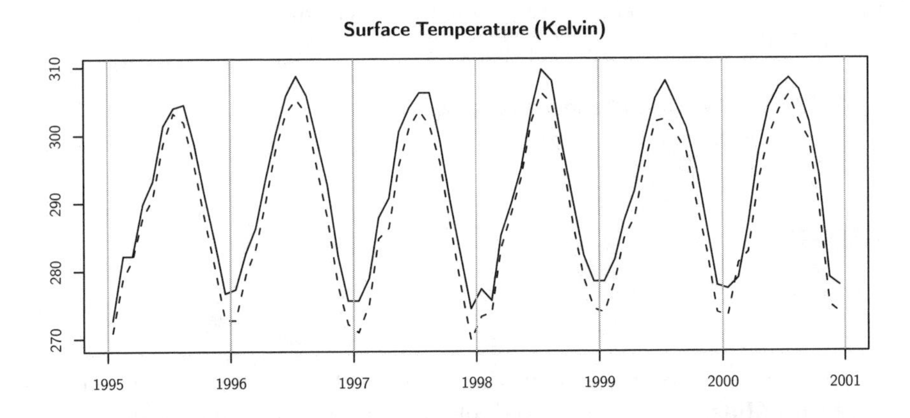

Figure 7.3: A plot of the surface temperature measurements from the 2006 JSM Data Expo, for all time points, at locations `1` (solid line) and `2` (dashed line). Vertical grey bars mark the change of years.

```
location  date  surftemp
--------  ----  --------
1         1     272.7
2         1     270.9
1         2     282.2
2         2     278.9
1         3     282.2
2         3     281.6
...
```

Figure 7.3 shows a plot of all of the values, which shows a clear trend of lower temperatures overall for location `2` (the dashed line).

The above query demonstrates that SQL code, even for a single query, can become quite long. This means that we should again apply the concepts from Section 2.4.3 to ensure that our code is tidy and easy to read. The code in this chapter provides many examples of the use of indenting to maintain order within an SQL query.

As well as extracting raw values from a column, it is possible to calculate derived values by combining columns with simple arithmetic operators or by using a **function** to produce the sum or average of the values in a column.

As a simple example, the following code calculates the average surface temperature value across all locations and across all time points. It crudely represents the average surface temperature of Central America for the years 1995 to 2000.

```
SQL> SELECT AVG(surftemp) avgstemp
         FROM measure_table;
```

```
avgstemp
--------
296.2310
```

One extra feature to notice about this example SQL query is that it defines a **column alias**, `avgstemp`, for the column of averages. The components of this part of the code are shown below.

keyword:	<u>`SELECT`</u> `AVG(surftemp) avgstemp`
column name:	`SELECT` <u>**`AVG(surftemp)`**</u> `avgstemp`
column alias:	`SELECT AVG(surftemp)` <u>**`avgstemp`**</u>

This part of the query specifies that we want to select the average surface temperature value, `AVG(surftemp)`, *and* that we want to be able to refer to this column by the name `avgstemp`. This alias can be used within the SQL query, which can make the query easier to type and easier to read, and the alias is also used in the presentation of the result. Column aliases will become more important as we construct more complex queries later in the section.

An SQL function will produce a single overall value for a column of a table, but what is usually more interesting is the value of the function for subgroups within a column, so the use of functions is commonly combined with a `GROUP BY` clause, which results in a separate summary value computed for subsets of the column.

For example, instead of investigating the trend in surface temperature over time for just location `1`, we could look at the change in the surface temperature over time averaged across all locations (i.e., the average surface temperature for each month).

The following code performs this query and Figure 7.4 shows a plot of the result. The `GROUP BY` clause specifies that we want an average surface temperature value for each different value in the `date` column (i.e., for each month).

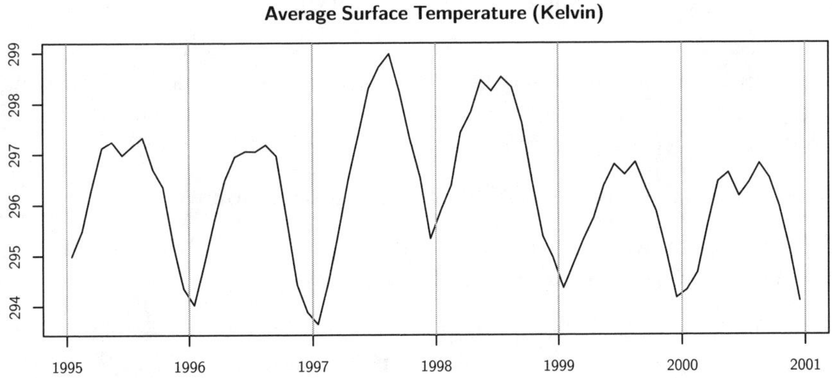

Figure 7.4: A plot of the surface temperature measurements from the 2006 JSM Data Expo, averaged across all locations, for each time point. Vertical grey bars mark the change of years.

```
SQL> SELECT date, AVG(surftemp) avgstemp
         FROM measure_table
         GROUP BY date
         ORDER BY date;

date  avgstemp
----  --------
1     294.9855
2     295.4869
3     296.3156
4     297.1197
5     297.2447
6     296.9769
...
```

Overall, it appears that 1997 and 1998 were generally warmer years in Central America. This result probably corresponds to the major El Niño event of 1997–1998.

7.2.3 Collations

There can be ambiguity whenever we sort or compare text values. A simple example of this issue is deciding whether an upper-case 'A' comes before a lower-case 'a'. More complex issues arise when comparing text from different languages.

The solution to this ambiguity is to explicitly specify a rule for comparing or sorting text. For example, a case-insensitive rule says that 'A' and 'a' should be treated as the same character.

In most databases, this sort of rule is called a **collation**.

Unfortunately, the default collation that is used may differ between database systems, as can the syntax for specifying a collation.

For example, with SQLite, the default is to treat text as case-sensitive, and a case-insensitive ordering can be obtained by adding a `COLLATE NOCASE` clause to a query.

In MySQL, it may be necessary to specify a collation clause, for example, `COLLATE latin1_bin`, in order to get case-sensitive sorting and comparisons.

7.2.4 Querying several tables: Joins

As demonstrated in the previous section, database queries from a single table are quite straightforward. However, most databases consist of more than one table, and most interesting database queries involve extracting information from more than one table. In database terminology, most queries involve some sort of **join** between two or more tables.

In order to demonstrate the most basic kind of join, we will briefly look at a new example data set.

7.2.5 Case study: Commonwealth swimming

The Commonwealth of Nations ("The Commonwealth") is a collection of 53 countries, most of which are former British colonies.

New Zealand sent a team of 18 swimmers to the Melbourne 2006 Commonwealth Games. Information about the swimmers, the events they competed in, and the results of their races are shown in Figure 7.5.

```
first   last   length  stroke        gender stage time    place
------  -----  ------  ------------  ------ ----- ------  -----
Zoe     Baker      50  Breaststroke  female heat   31.7       4
Zoe     Baker      50  Breaststroke  female semi   31.84      5
Zoe     Baker      50  Breaststroke  female final  31.45      4
Lauren  Boyle     200  Freestyle     female heat  121.11      8
Lauren  Boyle     200  Freestyle     female semi  120.9       8
Lauren  Boyle     100  Freestyle     female heat   56.7      10
Lauren  Boyle     100  Freestyle     female semi   56.4       9
...
```

Figure 7.5: A subset of the data recorded for New Zealand swimmers at the Melbourne 2006 Commonwealth Games, including the name and gender of each swimmer and the distance, stroke, stage, and result for each event that they competed in.

These data have been stored in a database with six tables.

The `swimmer_table` has one row for each swimmer and contains the first and last name of each swimmer. Each swimmer also has a unique numeric identifier.

```
swimmer_table ( ID [PK], first, last )
```

There are four tables that define the set of valid events: the distances are `50m`, `100m`, and `200m`; the swim strokes are breaststroke (`Br`), freestyle (`Fr`), butterfly (`Bu`), and backstroke (`Ba`); the genders are male (`M`) and female (`F`); and the possible race stages are heats (`heat`), semifinals (`semi`), and finals (`final`).

```
distance_table ( length [PK] )
stroke_table ( ID [PK], stroke )
gender_table ( ID [PK], gender )
stage_table ( stage [PK] )
```

The `result_table` contains information on the races swum by individual swimmers. Each row specifies a swimmer and the type of race (distance, stroke, gender, and stage). In addition, the swimmer's time and position in the race (`place`) are recorded.

```
result_table ( swimmer [PK] [FK swimmer_table.ID],
               distance [PK] [FK distance_table.length],
               stroke [PK] [FK stroke_table.ID],
               gender [PK] [FK gender_table.ID],
               stage [PK] [FK stage_table.stage],
               time, place )
```

The database design is illustrated in the diagram below.

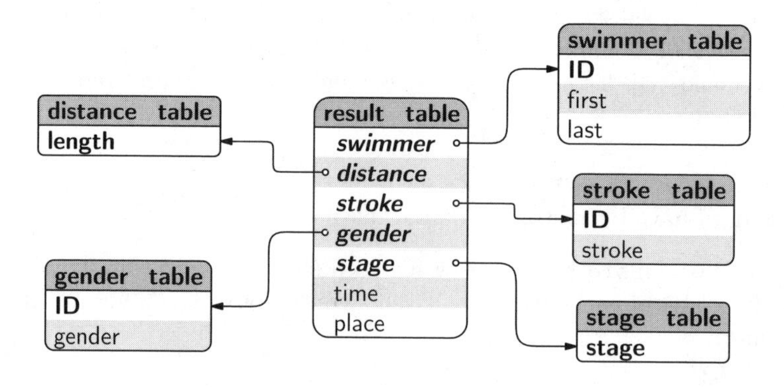

As an example of the information stored in this database, the following code shows that the swimmer with an ID of 1 is called Zoe Baker.

```
SQL> SELECT * FROM swimmer_table
        WHERE ID = 1;


ID  first  last
--  -----  -----
1   Zoe    Baker
```

Notice the use of * in this query to denote that we want *all* columns from the table in our result.

The following code shows that Zoe Baker swam in three races—a heat, a semifinal and the final of the women's 50m breaststroke—and she came 4th in the final in a time of 31.45 seconds.

```
SQL> SELECT * FROM result_table
       WHERE swimmer = 1;

swimmer distance stroke gender stage time  place
------- -------- ------ ------ ----- ----- -----
1       50       Br     F      final 31.45 4
1       50       Br     F      heat  31.7  4
1       50       Br     F      semi  31.84 5
```

7.2.6 Cross joins

The most basic type of database join, upon which all other types of join are conceptually based, is a **cross join**. The result of a cross join is the Cartesian product of the rows of one table with the rows of another table. In other words, row 1 of table 1 is paired with each row of table 2, then row 2 of table 1 is paired with each row of table 2, and so on. If the first table has n_1 rows and the second table has n_2 rows, the result of a cross join is a table with $n_1 \times n_2$ rows.

The simplest way to create a cross join is simply to perform an SQL query on more than one table. As an example, we will perform a cross join on the `distance_table` and `stroke_table` in the swimming database to generate all possible combinations of swimming stroke and event distance. The `distance_table` has three rows.

```
SQL> SELECT *
       FROM distance_table;

length
------
50
100
200
```

The `stroke_table` has four rows.

```
SQL> SELECT *
       FROM stroke_table;

ID stroke
-- ------------
Br Breaststroke
Fr Freestyle
Bu Butterfly
Ba Backstroke
```

A cross join between these tables has 12 rows, including all possible combinations of the rows of the two tables.

```
SQL> SELECT length, stroke
         FROM distance_table, stroke_table;


length  stroke
------  ------------
50      Breaststroke
50      Freestyle
50      Butterfly
50      Backstroke
100     Breaststroke
100     Freestyle
100     Butterfly
100     Backstroke
200     Breaststroke
200     Freestyle
200     Butterfly
200     Backstroke
```

A cross join can also be obtained more explicitly using the CROSS JOIN syntax as shown below (the result is exactly the same as for the code above).

```
SELECT length, stroke
    FROM distance_table CROSS JOIN stroke_table;
```

We will come back to this data set later in the chapter.

7.2.7 Inner joins

An **inner join** is the most common way of combining two tables. In this sort of join, only "matching" rows are extracted from two tables. Typically, a foreign key in one table is matched to the primary key in another table.

This is the natural way to combine information from two separate tables.

Conceptually, an inner join is a cross join, but with only the desired rows retained.

In order to demonstrate inner joins, we will return to the Data Expo database (see Section 7.1).

7.2.8 Case study: The Data Expo (continued)

In a previous example (page 163), we saw that the surface temperatures from the Data Expo data set for location 1 were consistently higher than the surface temperatures for location 2. Why is this?

One obvious possibility is that location 1 is closer to the equator than location 2. To test this hypothesis, we will repeat the earlier query but add information about the latitude and longitude of the two locations.

To do this we need information from two tables. The surface temperatures come from the measure_table and the longitude/latitude information comes from the location_table.

The following code performs an inner join between these two tables, combining rows from the measure_table with rows from the location_table that have the same location ID.

```
SQL> SELECT longitude, latitude, location, date, surftemp
         FROM measure_table mt, location_table lt
         WHERE location = ID AND
               (location = 1 OR
                location = 2)
         ORDER BY date;
```

```
longitude  latitude  location  date  surftemp
---------  --------  --------  ----  --------
-113.75    36.25     1         1     272.7
-111.25    36.25     2         1     270.9
-113.75    36.25     1         2     282.2
-111.25    36.25     2         2     278.9
-113.75    36.25     1         3     282.2
-111.25    36.25     2         3     281.6
...
```

The result shows that the longitude for location 2 is less negative (less westward) than the longitude for location 1, so the difference between the locations is that location 2 is to the east of location 1 (further inland in the US southwest).

The most important feature of this code is the fact that it obtains information from two tables.

```
FROM measure_table mt, location_table lt
```

Another important feature of this code is that it makes use of **table aliases**. The components of this part of the code are shown below.

```
     keyword:  FROM measure_table mt, location_table lt
  table name:  FROM measure_table mt, location_table lt
 table alias:  FROM measure_table mt, location_table lt
  table name:  FROM measure_table mt, location_table lt
 table alias:  FROM measure_table mt, location_table lt
```

We have specified that we want information from the `measure_table` *and* we have specified that we want to use the alias `mt` to refer to this table within the code of this query. Similarly, we have specified that the alias `lt` can be used instead of the full name `location_table` within the code of this query. This makes it easier to type the code and can also make it easier to read the code.

A third important feature of this code is that, unlike the cross join from the previous section, in this join we have specified that the rows from one table must match up with rows from the other table. In most inner joins, this means specifying that a foreign key from one table matches the primary key in the other table, which is precisely what has been done in this case; the `location` column from the `measure_table` is a foreign key that references the `ID` column from the `location_table`.

```
WHERE location = ID
```

The result is that we get the longitude and latitude information combined with the surface temperature information for the same location.

The `WHERE` clause in this query also demonstrates the combination of three separate conditions: there is a condition matching the foreign key of the `measure_table` to the primary key of the `location_table`, *plus* there are two conditions that limit our attention to just two values of the `location` column. The use of parentheses is important to control the order in which the conditions are combined.

Another way to specify the join in the previous query uses a different syntax that places all of the information about the join in the `FROM` clause of the query. The following code produces exactly the same result as before but uses the key words `INNER JOIN` between the tables that are being joined and follows that with a specification of the columns to match `ON`. Notice how the `WHERE` clause is much simpler in this case.

```
SELECT longitude, latitude, location, date, surftemp
    FROM measure_table mt
        INNER JOIN location_table lt
            ON mt.location = lt.ID
    WHERE location = 1 OR
          location = 2
    ORDER BY date;
```

This idea of joining tables extends to more than two tables. In order to demonstrate this, we will now consider a major summary of temperature values: what is the average temperature *per year*, across all locations *on land* (above sea level)?

In order to answer this question, we need to know the temperatures from the `measure_table`, the elevation from the `location_table`, and the years from the `date_table`. In other words, we need to combine all three tables.

This situation is one reason for using the `INNER JOIN` syntax shown above, because it naturally extends to joining more than two tables and results in a clearer and tidier query. The following code performs the desired query (see Figure 7.6).

```
SQL> SELECT year, AVG(surftemp) avgstemp
         FROM measure_table mt
              INNER JOIN location_table lt
                  ON mt.location = lt.ID
              INNER JOIN date_table dt
                  ON mt.date = dt.ID
         WHERE elevation > 0
         GROUP BY year;


year  avgstemp
----  --------
1995  295.3807
1996  295.0065
1997  295.3839
1998  296.4164
1999  295.2584
2000  295.3150
```

The result in Figure 7.6 shows only 1998 as warmer than other years, which suggests that the higher temperatures for 1997 that we saw in Figure 7.4 were due to higher temperatures over water.

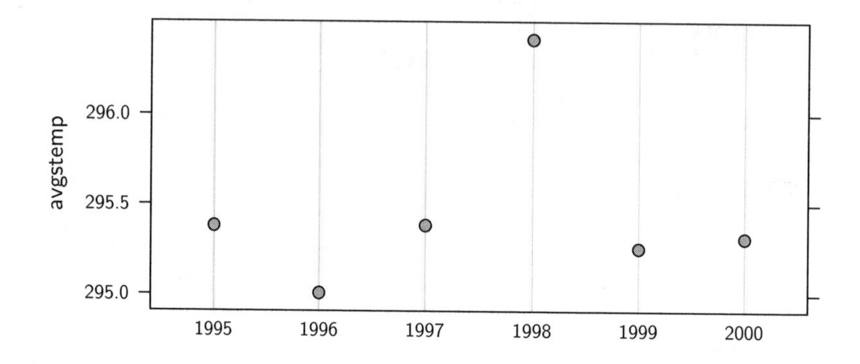

Figure 7.6: A plot of the surface temperature measurements from the 2006 JSM Data Expo, averaged across all locations with an elevation greater than zero and averaged across months, for each year.

There is another important new feature of SQL syntax in the code for this query, which occurs within the part of the code that specifies which columns the inner join should match `ON`. This part of the code is reproduced below.

```
ON mt.location = lt.ID
```

This code demonstrates that, within an SQL query, a column may be specified using a combination of the table name and the column name, rather than just using the column name. In this case, we have defined an alias for each table, so we can use the table alias rather than the complete table name. The components of this part of the code are shown below.

table name (alias): `ON mt.location = lt.ID` (mt)
column name: `ON mt.location = lt.ID` (location)
table name (alias): `ON mt.location = lt.ID` (lt)
column name: `ON mt.location = lt.ID` (ID)

This syntax is important when joining several tables because the same column name can be used in two different tables. This is the case in the above example; both the `location_table` and the `date_table` have a column called `ID`. This syntax allows us to specify exactly which `ID` column we mean.

7.2.9 Subqueries

It is possible to use an SQL query within another SQL query, in which case the nested query is called a **subquery**.

As a simple example, consider the problem of extracting the date at which the lowest surface temperature occurred. It is simple enough to determine the minimum surface temperature.

```
SQL> SELECT MIN(surftemp) min FROM measure_table;

min
-----
266.0
```

In order to determine the date, we need to find the row of the measurement table that matches this minimum value. We can do this using a subquery as shown below.

```
SQL> SELECT date, surftemp stemp
         FROM measure_table
         WHERE surftemp = ( SELECT MIN(surftemp)
                                FROM measure_table );

date  stemp
----  -----
36    266.0
```

The query that calculates the minimum surface temperature is inserted within parentheses as a subquery within the WHERE clause. The outer query returns only the rows of the measure_table where the surface temperature is equal to the minimum.

This subquery can be part of a more complex query. For example, the following code also performs a join so that we can see the real dates that these maximum temperatures occurred.

```
SQL> SELECT year, month, surftemp stemp
         FROM measure_table mt
             INNER JOIN date_table dt
                 ON mt.date = dt.ID
         WHERE surftemp = ( SELECT MIN(surftemp)
                                FROM measure_table );

year  month     stemp
----  --------  -----
1997  December  266.0
```

7.2.10 Outer joins

Another type of table join is the **outer join**, which differs from an inner join by including additional rows in the result.

In order to demonstrate this sort of join, we will return to the Commonwealth swimming example.

7.2.11 Case study: Commonwealth swimming (continued)

The results of New Zealand's swimmers at the 2006 Commonwealth Games in Melbourne are stored in a database consisting of six tables: a table of information about each swimmer; separate tables for the distance of a swim event, the type of swim stroke, the gender of the swimmers in an event, and the stage of the event (heat, semifinal, or final); plus a table of results for each swimmer in different events.

In Section 7.2.6 we saw how to generate all possible combinations of distance and stroke in the swimming database using a cross join between the `distance_table` and the `stroke_table`. There are three possible distances and four different strokes, so the cross join produced 12 different combinations.

We will now take that cross join and combine it with the table of race results using an inner join.

Our goal is to summarize the result of all races for a particular combination of distance and stroke by calculating the average time from such races. The following code performs this inner join, with the results ordered from fastest event on average to slowest event on average.

The cross join produces all possible combinations of distance and stroke and the result table is then joined to that, making sure that the results match up with the correct distance and stroke.

```
SQL> SELECT dt.length length,
            st.stroke stroke,
            AVG(time) avg
         FROM distance_table dt
             CROSS JOIN stroke_table st
             INNER JOIN result_table rt
                 ON dt.length = rt.distance AND
                    st.ID = rt.stroke
         GROUP BY dt.length, st.stroke
         ORDER BY avg;
```

```
length  stroke        avg
------  ------------  -----
50      Freestyle     26.16
50      Butterfly     26.40
50      Backstroke    28.04
50      Breaststroke  31.29
100     Butterfly     56.65
100     Freestyle     57.10
100     Backstroke    60.55
100     Breaststroke  66.07
200     Freestyle     118.6
200     Butterfly     119.0
200     Backstroke    129.7
```

The result suggests that freestyle and butterfly events tend to be faster on average than breaststroke and backstroke events.

However, the feature of the result that we need to focus on for the current purpose is that this result has only *11* rows.

What has happened to the remaining combination of distance and stroke? The answer is that, for inner joins, a row is not included in the result if either of the two columns being matched in the `ON` clause has the value `NULL`.

In this case, one row from the cross join, which produced all possible combinations of distance and stroke, has been dropped from the result because this combination does not appear in the `result_table`; no New Zealand swimmer competed in the 200m breaststroke.

This feature of inner joins is not always desirable and can produce misleading results, which is why an **outer join** is sometimes necessary. The idea of an outer join is to retain in the final result rows where one or another of the columns being matched has a NULL value.

The following code repeats the previous query, but instead of using INNER JOIN, it uses LEFT JOIN to perform a **left outer join** so that all distance/stroke combinations are reported, even though there is no average time information available for one of the combinations. The result now includes all possible combinations of distance and stroke, with a NULL value where there is no matching avg value from the result_table.

```
SQL> SELECT dt.length length,
            st.stroke stroke,
            AVG(time) avg
         FROM distance_table dt
             CROSS JOIN stroke_table st
             LEFT JOIN result_table rt
                 ON dt.length = rt.distance AND
                    st.ID = rt.stroke
         GROUP BY dt.length, st.stroke
         ORDER BY avg;
```

```
length  stroke        avg
------  ------------  -----
200     Breaststroke  NULL
50      Freestyle     26.16
50      Butterfly     26.40
50      Backstroke    28.04
50      Breaststroke  31.29
100     Butterfly     56.65
100     Freestyle     57.10
100     Backstroke    60.55
100     Breaststroke  66.07
200     Freestyle     118.6
200     Butterfly     119.0
200     Backstroke    129.7
```

The use of LEFT JOIN in this example is significant because it means that all rows from the original cross join are retained even if there is no matching row in the result_table.

It is also possible to use `RIGHT JOIN` to perform a **right outer join** instead. In that case, all rows of the `result_table` (the table on the right of the join) would have been retained.

In this case, the result of a right outer join would be the same as using `INNER JOIN` because all rows of the `result_table` have a match in the cross join. This is not surprising because it is equivalent to saying that all swimming results came from events that are a subset of all possible combinations of event stroke and event distance.

It is also possible to use `FULL JOIN` to perform a **full outer join**, in which case all rows from tables on *both* sides of the join are retained in the final result.

7.2.12 Self joins

It is useful to remember that database joins always begin, at least conceptually, with a Cartesian product of the rows of the tables being joined. The different sorts of database joins are all just different subsets of a cross join. This makes it possible to answer questions that, at first sight, may not appear to be database queries.

For example, it is possible to join a table with itself in what is called a **self join**, which produces all possible combinations of the rows of a table. This sort of join can be used to answer questions that require comparing a column within a table to itself or to other columns within the same table. The following case study provides an example.

7.2.13 Case study: The Data Expo (continued)

Consider the following question: did the temperature at location 1 for January 1995 (date 1) occur at any *other* locations and times?

This question requires a comparison of one row of the `temperature` column in the `measure_table` with the other rows in that column. The code below performs the query using a self join.

```
SQL> SELECT mt1.temperature temp1, mt2.temperature temp2,
            mt2.location loc, mt2.date date
         FROM measure_table mt1, measure_table mt2
         WHERE mt1.temperature = mt2.temperature AND
               mt1.date = 1 AND
               mt1.location = 1;

temp1  temp2  loc  date
-----  -----  ---  ----
272.1  272.1  1    1
272.1  272.1  498  13
```

To show the dates as real dates and locations as longitudes and latitudes, we can join the result to the `date_table` as well.

```
SQL> SELECT mt1.temperature temp1, mt2.temperature temp2,
            lt.longitude long, lt.latitude lat, dt.date date
         FROM measure_table mt1, measure_table mt2
            INNER JOIN date_table dt
               ON mt2.date = dt.ID
            INNER JOIN location_table lt
               ON mt2.location = lt.ID
         WHERE mt1.temperature = mt2.temperature AND
               mt1.date = 1 AND
               mt1.location = 1;


temp1  temp2  long     lat     date
-----  -----  -------  ------  ----------
272.1  272.1  -113.75  36.25   1995-01-16
272.1  272.1  -71.25   -13.75  1996-01-16
```

The temperature occurred again for January 1996 in a location far to the east and south of location 1.

7.2.14 Running SQL code

One major advantage of SQL is that it is implemented by every major DBMS. This means that we can learn a single language and then use it to work with any database.

Not all DBMS software supports all of the SQL standard and most DBMS software has special features that go beyond the SQL standard. However, the basic SQL queries that are described in this chapter should work in any major DBMS.

Another major difference between DBMS software is the user interface. In particular, the commercial systems tend to have complete GUIs while the open-source systems tend to default to a command-line interface. However, even a DBMS with a very sophisticated GUI will have a menu option somewhere that will allow SQL code to be run.

The simplest DBMS for experimenting with SQL is the SQLite system[1] because it is very straightforward to install. This section provides a very brief example SQLite session.

SQLite is run from a command window or shell by typing the name of the program plus the name of the database that we want to work with. For example, at the time of writing, the latest version of SQLite was named `sqlite3`. We would type the following to work with the `dataexpo` database.

```
sqlite3 dataexpo
```

SQLite then presents a prompt, usually `sqlite>`. We type SQL code after the prompt and the result of our code is printed out to the screen. For example, a simple SQL query with the `dataexpo` database is shown below, with the result shown below the code.

```
sqlite> SELECT * FROM date_table WHERE ID = 1;
```

```
1|1995-01-16|January|1995
```

There are a number of special SQLite commands that control how the SQLite program works. For example, the `.mode`, `.header`, and `.width` commands control how SQLite formats the results of queries. The following example shows the use of these special commands to make the result include column names and to use fixed-width columns for displaying results.

```
sqlite> .header ON
sqlite> .mode column
sqlite> .width 2 10 7 4
sqlite> SELECT * FROM date_table WHERE ID = 1;
```

```
ID  date        month    year
--  ----------  -------  ----
1   1995-01-16  January  1995
```

[1]`http://www.sqlite.org/`

The full set of these special SQLite commands can be viewed by typing `.help`.

To exit the SQLite command line, type `.exit`.

Recap

The SQL SELECT command is used to query (extract information from) a relational database.

An SQL query can limit which columns and which rows are returned from a database table.

Information can be combined from two or more database tables using some form of database join: a cross join, an inner join, an outer join, or a self join.

7.3 Querying XML

The direct counterpart to SQL that is designed for querying XML documents is a language called XQuery. However, a full discussion of XQuery is beyond the scope of this book. Instead, this section will only focus on XPath, a language that underpins XQuery, as well as a number of other XML-related technologies.

The XPath language provides a way to specify a particular subset of an XML document. XPath makes it easy to identify a coherent set of data values that are distributed across multiple elements within an XML document.

7.3.1 XPath syntax

An XPath **expression** specifies a subset of elements and attributes from within an XML document. We will look at the basic structure of XPath expressions via an example.

7.3.2 Case study: Point Nemo (continued)

Figure 7.7 shows the temperature data at Point Nemo in an XML format (this is a reproduction of Figure 5.13 for convenience).

```
<?xml version="1.0"?>
<temperatures>
    <variable>Mean TS from clear sky composite (kelvin)</variable>
    <filename>ISCCPMonthly_avg.nc</filename>
    <filepath>/usr/local/fer_dsets/data/</filepath>
    <subset>93 points (TIME)</subset>
    <longitude>123.8W(-123.8)</longitude>
    <latitude>48.8S</latitude>
    <case date="16-JAN-1994" temperature="278.9" />
    <case date="16-FEB-1994" temperature="280" />
    <case date="16-MAR-1994" temperature="278.9" />
    <case date="16-APR-1994" temperature="278.9" />
    <case date="16-MAY-1994" temperature="277.8" />
    <case date="16-JUN-1994" temperature="276.1" />

    ...

</temperatures>
```

Figure 7.7: The first few lines of the surface temperature at Point Nemo in an XML format. This is a reproduction of Figure 5.16.

This XML document demonstrates the idea that values from a single variable in a data set may be scattered across separate XML elements. For example, the temperature values are represented as attributes of `case` elements; they are not assembled together within a single column or a single block of memory within the file.

We will use some XPath expressions to extract useful subsets of the data set from this XML document.

The most basic XPath expressions consist of element names separated by forwardslashes. The following XPath selects the `temperatures` element from the XML document. In each of the XPath examples in this section, the elements or attributes that are selected by the XPath expression will be shown below the XPath code. If there are too many elements or attributes, then, to save space, only the first few will be shown, followed by ... to indicate that some of the results were left out.

```
/temperatures
```

```
<temperatures>
  <variable>Mean TS from clear sky composite (kelvin)</variable>
  <filename>ISCCPMonthly_avg.nc</filename>
 ...
```

More specifically, because the expression begins with a forwardslash, it selects the *root* element called `temperatures`. If we want to select elements below the root element, we need to specify a complete path to those elements, or start the expression with a double-forwardslash.

Both of the following two expressions select all `case` elements from the XML document. In the first example, we specify `case` elements that are directly nested within the (root) `temperatures` element:

```
/temperatures/case
```

```
<case date="16-JAN-1994" temperature="278.9"/>
<case date="16-FEB-1994" temperature="280"/>
<case date="16-MAR-1994" temperature="278.9"/>
 ...
```

The second approach selects `case` elements no matter where they are within the XML document.

```
//case
```

```
<case date="16-JAN-1994" temperature="278.9"/>
<case date="16-FEB-1994" temperature="280"/>
<case date="16-MAR-1994" temperature="278.9"/>
 ...
```

An XPath expression may also be used to subset attributes rather than entire elements. Attributes are selected by specifying the appropriate name, preceded by an `@` character. The following example selects the `temperature` attribute from the `case` elements.

```
/temperatures/case/@temperature
```

```
temperature="278.9"
temperature="280"
temperature="278.9"
...
```

Several separate paths may also be specified, separated by a vertical bar. This next XPath selects both `longitude` and `latitude` elements from anywhere within the XML document.

```
//longitude | //latitude
```

```
<longitude>123.8W(-123.8)</longitude>
<latitude>48.8S</latitude>
```

It is also possible to specify **predicates**, which are conditions that must be met for an element to be selected. These are placed within square brackets. In the following example, only `case` elements where the `temperature` attribute has the value 280 are selected.

```
/temperatures/case[@temperature=280]
```

```
<case date="16-FEB-1994" temperature="280"/>
<case date="16-MAR-1995" temperature="280"/>
<case date="16-MAR-1997" temperature="280"/>
```

We will demonstrate more examples of the use of XPath expressions later in Section 9.7.7, which will include an example of software that can be used to run XPath code.

7.4 Further reading

The w3schools XPath Tutorial
    `http://www.w3schools.com/xpath/`
    Quick, basic tutorial-based introduction to XPath.

Summary

When data has been stored using a more sophisticated data storage format, a more sophisticated tool is required to access the data.

SQL is a language for accessing data that has been stored in a relational database.

XPath is a language for specifying a subset of data values in an XML document.

8

SQL Reference

The Structured Query Language (SQL) is a language for working with information that has been stored in a database.

SQL has three parts: the Data Manipulation Language (DML) concerns adding information to a database, modifying the information, and extracting information from a database; the Data Definition Language (DDL) is concerned with the structure of a database (creating tables); and the Data Control Language (DCL) is concerned with administration of a database (deciding who gets what sort of access to which parts of the database).

This chapter is mostly focused on the SELECT command, which is the part of the DML that is used to extract information from a database, but other useful SQL commands are also mentioned briefly in Section 8.3.

SQL

8.1 SQL syntax

This section is only concerned with the syntax of the SQL SELECT command, which is used to perform a database query.

Within this chapter, any code written in a *sans-serif oblique font* represents a general template; that part of the code will vary depending on the database in use and the names of the tables as well as the names of the columns within those tables.

8.2 SQL queries

The basic format of an SQL query is this:

```
SELECT columns
    FROM tables
    WHERE row_condition
    ORDER BY order_by_columns
```

The SQL keywords, SELECT, FROM, WHERE, and ORDER BY, are traditionally written in uppercase, though this is not necessary.

The names of tables and columns depend on the database being queried, but they should always start with a letter and only contain letters, digits, and the underscore character, '_'.

This will select the named *columns* from the specified *tables* and return all rows matching the *row_condition*.

The order of the rows in the result is based on the columns named in the *order_by_columns* clause.

8.2.1 Selecting columns

The special character * selects all columns; otherwise, only those columns named are included in the result. If more than one column name is given, the column names must be separated by commas.

```
SELECT *
    ...

SELECT colname
    ...

SELECT colname1, colname2
    ...
```

The column name may be followed by a **column alias**, which can then be used anywhere within the query in place of the original column name (e.g., in the WHERE clause).

```
SELECT colname colalias
    ...
```

If more than one table is included in the query, and the tables share a column with the same name, a column name must be preceded by the relevant table name, with a full stop in between.

```
SELECT tablename.colname
    ...
```

Functions and operators may be used to produce results that are calculated from the column. The set of functions that is provided varies widely between

DBMS, but the normal mathematical operators for addition, subtraction, multiplication, and division, plus a set of basic aggregation functions for maximum value (MAX), minimum value (MIN), summation (SUM), and arithmetic mean (AVG), should always be available.

```
SELECT MAX(colname)
    ...

SELECT colname1 + colname2
    ...
```

A column name can also be a constant value (number or text), in which case the value is replicated for every row of the result.

8.2.2 Specifying tables: The FROM clause

The FROM clause must contain at least one table and all columns that are referred to in the query must exist in at least one of the tables in the FROM clause.

If a single table is specified, then the result is all rows of that table, subject to any filtering applied by a WHERE clause.

```
SELECT colname
    FROM tablename
    ...
```

A table name may be followed by a **table alias**, which can be used in place of the original table name anywhere else in the query.

```
SELECT talias.colname
    FROM tablename talias
    ...
```

If two tables are specified, separated only by a comma, the result is all possible combinations of the rows of the two tables (a Cartesian product). This is known as a **cross join**.

```
SELECT ...
    FROM table1, table2
    ...
```

An **inner join** is created from a cross join by specifying a condition so that only rows that have matching values are returned (typically using a foreign key to match with a primary key). The condition may be specified within the WHERE clause (see Section 8.2.3), or as part of an INNER JOIN syntax as shown below.

```
SELECT ...
    FROM table1 INNER JOIN table2
        ON table1.primarykey = table2.foreignkey
    ...
```

An **outer join** extends the inner join by including in the result rows from one table that have no match in the other table. There are left outer joins (where rows are retained from the table named on the left of the join syntax), right outer joins, and full outer joins (where non-matching rows from both tables are retained).

```
SELECT ...
    FROM table1 LEFT OUTER JOIN table2
        ON table1.primarykey = table2.foreignkey
    ...
```

A **self join** is a join of a table with itself. This requires the use of table aliases.

```
SELECT ...
    FROM tablename alias1, tablename alias2
    ...
```

8.2.3 Selecting rows: The WHERE clause

By default, all rows from a table, or from a combination of tables, are returned. However, if the WHERE clause is used to specify a condition, then only rows matching that condition will be returned.

Conditions may involve any column from any table that is included in the query. Conditions usually involve a comparison between a column and a constant value, or a comparison between two columns.

A constant text value should be enclosed in single-quotes.

Valid comparison operators include: equality (=), greater-than or less-than (>, <), greater-than or equal-to or less-than or equal-to (>=, <=), and inequality (!= or <>).

```
SELECT ...
    FROM ...
    WHERE colname = 0;
```

```
SELECT ...
    FROM ...
    WHERE column1 > column2;
```

Complex conditions can be constructed by combining simple conditions with logical operators: AND, OR, and NOT. Parentheses should be used to make the order of evaluation explicit.

```
SELECT ...
    FROM ...
    WHERE column1 = 0 AND
          column2 != 0;
```

```
SELECT ...
    FROM ...
    WHERE NOT (stroke = 'Br' AND
               (distance = 50 OR
                distance = 100));
```

For the case where a column can match several possible values, the special IN keyword can be used to specify a range of valid values.

```
SELECT ...
    FROM ...
    WHERE column1 IN (value1, value2);
```

Comparison with text constants can be generalized to allow patterns using the special LIKE comparison operator. In this case, within the text constant, the underscore character, _, has a special meaning; it will match *any single character*. The percent character, %, is also special and it will match *any number* of characters *of any sort*.

```
SELECT ...
    FROM ...
    WHERE stroke LIKE '%stroke';
```

8.2.4 Sorting results: The ORDER BY clause

The order of the *columns* in the results of a query is based on the order of the column names in the query.

The order of the *rows* in a result is undetermined unless an `ORDER BY` clause is included in the query.

The `ORDER BY` clause consists of one or more column names. The rows are ordered according to the values in the named columns. The keyword `ASC` is used to indicate ascending order and `DESC` is used for descending order.

```
SELECT ...
    FROM ...
    ORDER BY columnname ASC;
```

The results can be ordered by the values in several columns simply by specifying several column names, separated by commas. The results are ordered by the values in the first column, but if several rows in the first column have the same value, those rows are ordered by the values in the second column.

```
SELECT ...
    FROM ...
    ORDER BY column1 ASC, column2 DESC;
```

8.2.5 Aggregating results: The GROUP BY clause

The aggregation functions `MAX`, `MIN`, `SUM`, and `AVG` (see Section 8.2.1) all return a single value from a column. If a `GROUP BY` clause is included in the query, aggregated values are reported for each unique value of the column specified in the `GROUP BY` clause.

```
SELECT column1, SUM(column2)
    FROM ...
    GROUP BY column1;
```

Results can be reported for combinations of unique values of several columns simply by naming several columns in the `GROUP BY` clause.

```
SELECT column1, column2, SUM(column3)
    FROM ...
    GROUP BY column1, column2;
```

The GROUP BY clause can include a HAVING sub-clause that works like the WHERE clause but operates on the rows of aggregated results rather than the original rows.

```
SELECT column1, SUM(column2) colalias
    FROM ...
    GROUP BY column1
        HAVING colalias > 0;
```

8.2.6 Subqueries

The result of an SQL query may be used as part of a larger query. The subquery is placed within parentheses but otherwise follows the same syntax as a normal query.

Subqueries can be used in place of table names within the FROM clause and to provide comparison values within a WHERE clause.

```
SELECT column1
    FROM table1
    WHERE column1 IN
        ( SELECT column2
              FROM table2
                  ... );
```

8.3 Other SQL commands

This section deals with SQL commands that perform other common useful actions on a database.

We start with entering the data into a database.

Creating a table proceeds in two steps: first we must define the schema or structure of the table and then we can load rows of values into the table.

8.3.1 Defining tables

A table schema is defined using the `CREATE` command.

```
CREATE TABLE tablename
    (col1name col1type,
     col2name col2type)
    column_constraints;
```

This command specifies the name of the table, the name of each column, and the data type to be stored in each column. A common variation is to add `NOT NULL` after the column data type to indicate that the value of the column can never be `NULL`. This must usually be specified for primary key columns.

The set of possible data types available depends on the DBMS being used, but some standard options are shown in Table 8.1.

The *column_constraints* are used to specify primary and foreign keys for the table.

```
CREATE TABLE table1
    (col1name col1type NOT NULL,
     col2name col2type)
    CONSTRAINT constraint1
        PRIMARY KEY (col1name)
    CONSTRAINT constraint2
        FOREIGN KEY (col2name)
        REFERENCES table2 (table2col);
```

The primary key constraint specifies which column or columns make up the primary key for the table. The foreign key constraint specifies which column or columns in *this* table act as a foreign key *and* the constraint specifies the table and the column *in that table* that the foreign key refers to.

Table 8.1: Some common SQL data types.

Type	Description
CHAR(*n*)	Fixed-length text (*n* characters)
VARCHAR(*n*)	Variable-length text (maximum *n* characters)
INTEGER	Whole number
REAL	Real number
DATE	Calendar date

As concrete examples, the code in Figure 8.1 shows the SQL code that was used to create the database tables `date_table`, `location_table`, and `measure_table` for the Data Expo case study in Section 7.1.

The primary key of the `date_table` is the `ID` column and the primary key of the `location_table` is its `ID` column. The (composite) primary key of the `measure_table` is a combination of the `location` and `date` columns. The `measure_table` also has two foreign keys: the `date` column acts as a foreign key, referring to the `ID` column of the `date_table`, and the `location` column also acts as a foreign key, referring to the `ID` column of the `location_table`.

SQL

8.3.2 Populating tables

Having generated the table schema, values are entered into the table using the `INSERT` command.

```
INSERT INTO table VALUES
    (value1, value2);
```

There should be as many values as there are columns in the table, with values separated from each other by commas. Text values should be enclosed within single-quotes.

Most DBMS software also provides a way to read data values into a table from an external (text) file. For example, in SQLite, the special `.import` command can be used to read values from an external text file.

```
CREATE TABLE date_table
    (ID    INTEGER NOT NULL,
     date  DATE,
     month CHAR(9),
     year  INTEGER,
     CONSTRAINT date_table_pk PRIMARY KEY (ID));
CREATE TABLE location_table
    (ID    INTEGER NOT NULL,
     longitude   REAL,
     latitude    REAL,
     elevation   REAL,
     CONSTRAINT location_table_pk PRIMARY KEY (ID));
CREATE TABLE measure_table
    (location    INTEGER NOT NULL,
     date        INTEGER NOT NULL,
     cloudhigh   REAL,
     cloudmid    REAL,
     cloudlow    REAL,
     ozone       REAL,
     pressure    REAL,
     surftemp    REAL,
     temperature REAL,
     CONSTRAINT measure_table_pk
         PRIMARY KEY (location, date),
     CONSTRAINT measure_date_table_fk
         FOREIGN KEY (date)
         REFERENCES date_table(ID),
     CONSTRAINT measure_location_table_fk
         FOREIGN KEY (location)
         REFERENCES location_table(ID));
```

Figure 8.1: The SQL code used to define the table schema for storing the Data Expo data set in a relational database (see Section 7.1).

8.3.3 Modifying data

Values in a database table may be modified using the UPDATE command.

```
UPDATE table
    SET column = value
    WHERE row_condition
```

The rows of the specified *column*, within the specified *table*, that satisfy the *row_condition*, will be changed to the new *value*.

8.3.4 Deleting data

The DELETE command can be used to remove specific rows from a table.

```
DELETE FROM table
    WHERE row_condition;
```

The DROP command can be used to completely remove not only the contents of a table but also the entire table schema so that the table no longer exists within the database.

```
DROP TABLE table;
```

In some DBMS, it is even possible to "drop" an entire database (and all of its tables).

```
DROP DATABASE database;
```

These commands should obviously be used with extreme caution.

8.4 Further reading

SQL: The Complete Reference
by James R. Groff and Paul N. Weinberg
2nd Edition (2002) McGraw-Hill.
Like the title says, a *complete* reference to SQL.

Using SQL
by Rafe Colburn
Special Edition (1999) Que.
Still a thorough treatment, but an easier read.

9

Data Processing

In previous chapters, we have encountered some useful tools for specific computing tasks. For example, in Chapter 7 we learned about SQL for extracting data from a relational database.

Given a data set that has been stored in a relational database, we could now write a piece of SQL code to extract data from that data set. But what if we want to extract data from the database *at a specific time of day*? What if we want to repeat that task *every day for a year*? SQL has no concept of *when* to execute a task.

In Chapter 5, we learned about the benefits of the XML language for storing data, but we also learned that it is a very verbose format. We could conceive of designing the structure of an XML document, but would we really be prepared to write large amounts of XML *by hand*? What we want to be able to do is to get the computer to *write the file of XML code* for us.

In order to perform these sorts of tasks, we need a **programming language**. With a programming language, we will be able to tell the computer to perform a task at a certain time, or to repeat a task a certain number of times. We will be able to create files of information and we will be able to perform calculations with data values.

The purpose of this chapter is to explore and enumerate some of the tasks that a programming language will allow us to perform. As with previous topics, it is important to be aware of what tasks are actually possible as well as look at the technical details of how to carry out each task.

As we might expect, a programming language will let us do a lot more than the specific languages like HTML, XML, and SQL can do, but this will come at a cost because we will need to learn a few more complex concepts. However, the potential benefit is limitless. This is the chapter where we truly realize the promise of taking control of our computing environment.

The other important purpose of this chapter is to introduce a specific programming language, so that we can perform tasks in practice.

There are many programming languages to choose from, but in this chapter we will use the R language because it is *relatively* simple to learn and because it is particularly well suited to working with data.

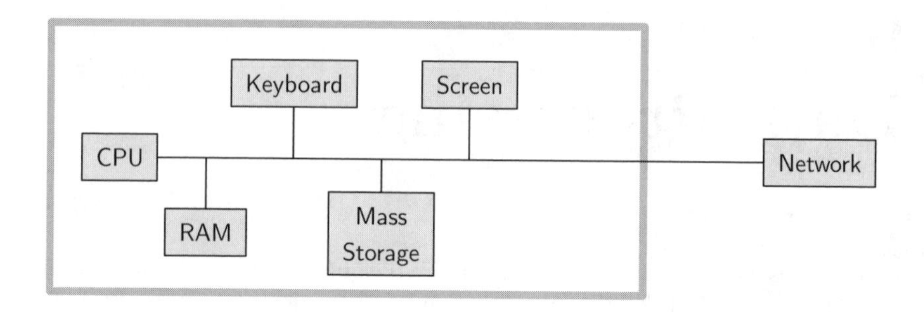

Figure 9.1: A diagram illustrating the basic components of a standard computing environment.

Computer hardware

So, what can we do with a programming language?

To answer that question, it will again be useful to compare and contrast a programming language with the more specific languages that we have already met and what we can do with those.

SQL lets us talk to database systems; we can ask database software to extract specific data from database tables. With HTML, we can talk to web browsers; we can instruct the browser software to draw certain content on a web page. Working with XML is a bit more promiscuous because we are essentially speaking to any software system that might be interested in our data. However, we are limited to only being able to say "here are the data".

A programming language is more general and more powerful than any of these, and the main reason for this is because a programming language allows us to talk not just to other software systems, but also to the computer *hardware*.

In order to understand the significance of this, we need to have a very basic understanding of the fundamental components of computer hardware and what we can do with them. Figure 9.1 shows a simple diagram of the basic components in a standard computing environment, and each component is briefly described below.

CPU

The Central Processing Unit (CPU) is the part of the hardware that can perform calculations or **process** our data. It is only capable of a small set of operations—basic arithmetic, plus the ability to compare values, and the ability to shift values between locations in computer memory—but it performs these tasks exceptionally quickly. Complex tasks are performed by combining many, many simple operations in the CPU.

The CPU also has access to a clock for determining times.

Being able to talk to the CPU means that we can perform arbitrarily complex calculations. This starts with simple things like determining minima, maxima, and averages for numeric values, but includes sophisticated data analysis techniques, and there really is no upper bound.

RAM

Computer **memory**, the hardware where data values are stored, comes in many different forms. Random Access Memory (RAM) is the term usually used to describe the memory that sits closest to the CPU. This memory is temporary—data values in RAM only last while the computer is running (they disappear when the computer is turned off)—and it is fast—values can be transferred to the CPU for processing and the result can be transferred back to RAM very rapidly. RAM is also usually *relatively* small.

Loosely speaking, RAM corresponds to the popular notion of short-term memory.

All processing of data typically involves, at a minimum, both RAM and the CPU. Data values are stored temporarily in RAM, shifted to the CPU to perform arithmetic or comparisons or something more complex, and the result is then stored back in RAM. A fundamental feature of a programming language is the ability to store values in RAM and specify the calculations that should be carried out by the CPU.

Being able to store data in RAM means that we can accomplish a complex task by performing a series of simpler steps. After each step, we record the intermediate result in memory so that we can use that result in subsequent steps.

Keyboard

The keyboard is one example of **input** hardware.

Most computers also have a mouse or touchpad. These are also examples of input hardware, but for the purposes of this book we are mostly interested in being able to enter code and data via a keyboard.

We will not be writing code to control the keyboard; this hardware component is more relevant to us as the primary way in which we will communicate our instructions and data to the computer.

Mass Storage

Most computers typically have a very large repository of computer memory, such as a hard drive. This will be much larger and much slower than RAM but has the significant advantage that data values will persist when the power goes off. Mass storage is where we save all of our files and documents.

A related set of hardware components includes external storage devices, such as CD and DVD drives and thumb drives (memory sticks), which allow the data to be physically transported away from the machine.

Where RAM corresponds to short-term memory, mass storage corresponds to long-term memory.

It is essential to be able to access mass storage because that is where the original data values will normally reside. With access to mass storage, we can also permanently store the results of our data processing as well as our computer code.

Screen

The computer screen is one example of **output** hardware. The screen is important as the place where text and images are displayed to show the results of our calculations.

Being able to control what is displayed on screen is important for viewing the results of our calculations and possibly for sharing those results with others.

Network

Most modern computers are connected to a network of some kind, which consists of other computers, printers, etc, and in many cases the general internet.

As with mass storage, the importance of having access to the network is that this may be where the original data values reside.

A programming language will allow us to work with these hardware components to perform all manner of useful tasks, such as reading files or documents from a hard disk into RAM, calculating new values, and displaying those new values on the computer screen.

How this chapter is organized

This chapter begins with a task that we might naturally think to perform by hand, but which can be carried out much more efficiently and accurately if instead we write code in a programming language to perform the task. The aims of this section are to show how useful a little programming knowledge can be and to demonstrate how an overall task can be broken down into smaller tasks that we can perform by writing code.

Sections 9.2 and 9.3 provide an initial introduction to the R programming language for performing these sorts of tasks. These sections will allow us to write and run some very basic R code.

Section 9.4 introduces the important idea of **data structures**—how data values are stored in RAM. In this section, we will learn how to enter data values by hand and how those values can be organized. All data processing tasks require data values to be loaded into RAM before we can perform calculations on the data values. Different processing tasks will require the data values to be organized and stored in different ways, so it is important to understand what options exist for storing data in RAM and how to work with each of the available options.

Some additional details about data structures are provided in Section 9.6, but before that, Section 9.5 provides a look at one of the most basic data processing tasks, which is extracting a **subset** from a large set of values. Being able to break a large data set into smaller pieces is one of the fundamental small steps that we can perform with a programming language. Solutions to more complex tasks are based on combinations of these small steps.

Section 9.7 addresses how to get data from external files into RAM so that we can process them. Most data sets are stored permanently in some form of mass storage, so it is crucial to know how to load data from various storage formats into RAM using R.

Section 9.8 describes a number of data processing tasks. Much of the chapter up to this point is laying the foundation. This section starts to provide information for performing powerful calculations with data values. Again, the individual techniques that we learn in these sections provide the foundation for completing more complex tasks. Section 9.8.12 provides a larger case study that demonstrates how the smaller steps can be combined to carry out a more substantial data processing exercise.

Section 9.9 looks at the special case of processing text data. We will look at tools for searching within text, extracting subsets from text, and splitting and recombining text values. Section 9.9.2 describes **regular expressions**, which are an important tool for searching for patterns within text.

Section 9.10 describes how to format the results of data processing, either for display on screen or for use in research reports.

Section 9.11 very briefly discusses some more advanced ideas about writing code in a programming language, and Section 10 contains a few comments about and pointers to alternative software and programming languages for data processing. These are more advanced topics and provide a small glimpse of areas to explore further.

9.1 Case study: The Population Clock

The Doomsday Clock symbolizes how close the world is to complete disaster. It currently stands at 5 minutes to midnight ...

The U.S. Census Bureau maintains a web site called the World Population Clock (see Figure 9.2).

This web site provides an up-to-the-minute snapshot of the world's population, based on estimates by the U.S. Census Bureau. It is updated every few seconds.

In this case study, we will use this clock to generate a rough estimate of the current *rate of growth* of the world's population.

We will perform this task by taking two snapshots of the World Population Clock, ten minutes apart, and then we will divide the change in population by the change in time.

The purpose of this section is to look at the steps involved in detail, noting how we might perform this task "by hand", and then looking at how we might use the computer to do the work instead. This will allow us to see what sorts of tasks we can expect a computer to be able to perform and will

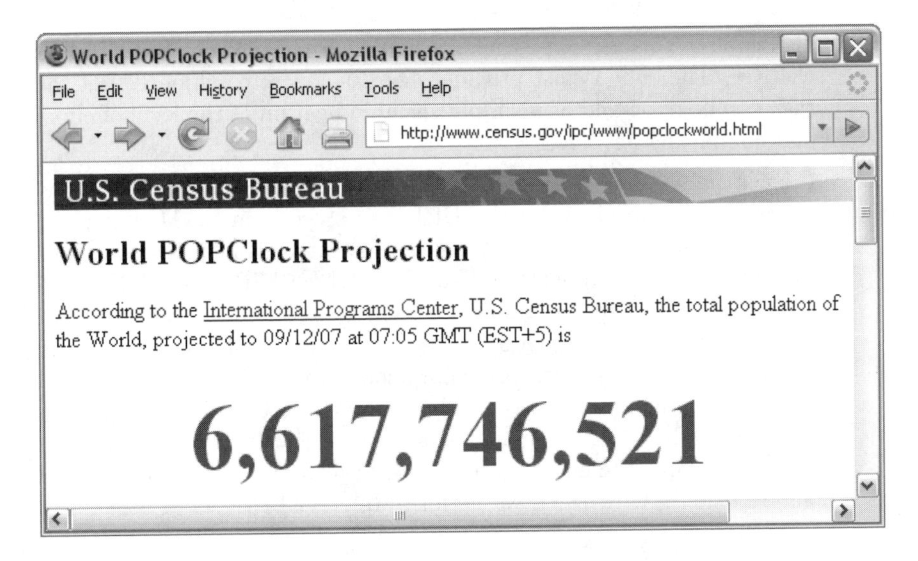

Figure 9.2: The World Population Clock web page shows an up-to-the-minute snapshot of the world's population (based on estimates by the U.S. Census Bureau).

begin to introduce some of the programming concepts involved.

1. **Copy the current value of the population clock.**

   The first step is to capture a snapshot of the world population from the U.S. Census Bureau web site.

   This is very easy to do by hand; we simply navigate a web browser to the population clock web page and type out, cut-and-paste, or even just write down the current population value.

   What about getting the computer to do the work?

   Navigating to a web page and downloading the information is not actually very difficult. This is an example of interacting with the network component of the computing environment. Downloading a web page is something that we can expect any modern programming language to be able to do, given the appropriate URL (which is visible in the "navigation bar" of the web browser in Figure 9.2).

   The following R code will perform this **data import** task.

   ```
   R> clockHTML <-
        readLines("http://www.census.gov/ipc/www/popclockworld.html")
   ```

   We will not focus on understanding all of the details of the examples of R code in this section—that is the purpose of the remainder of this chapter. The code is just provided here as concrete evidence that the task can be done and as a simple visual indication of the level of effort and complexity involved.

Conceptually, the above code says "read the HTML code from the network location given by the URL and store it in RAM under the name `clockHTML`." The images below illustrate this idea, showing how the information that we input at the keyboard (the URL) leads to the location of a file containing HTML code on the network, which is read into RAM and given the name `clockHTML`. The image on the left shows the main hardware components involved in this process in general and the image on the right shows the actual data values and files involved in this particular example. We will use diagrams like this throughout the chapter to illustrate which hardware components we are dealing with when we perform different tasks.

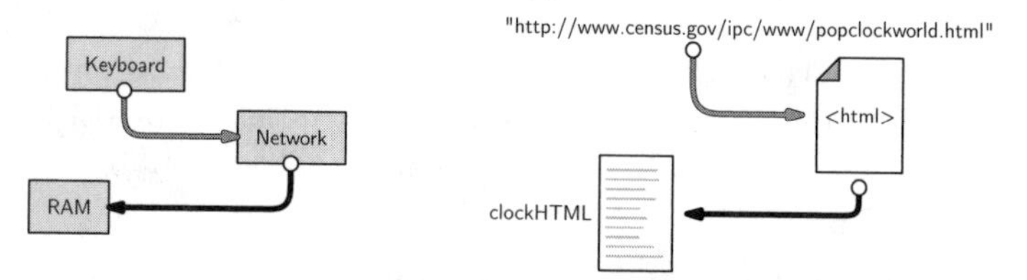

It is important to realize that the result of the R code above is *not* a nice picture of the web page like we see in a browser. Instead, we have the raw HTML code that describes the web page (see Figure 9.3).

This is actually a good thing because it would be incredibly difficult for the computer to extract the population information from a picture.

The HTML code is better than a picture because the HTML code has a clear structure. If information has a pattern or structure, it is much easier to write computer code to navigate within the information. We will exploit the structure in the HTML code to get the computer to extract the relevant population value for us.

However, before we do anything with this HTML code, it is worth taking note of what sort of information we have. From Chapter 2, we know that HTML code is just plain text, so what we have downloaded is a plain text file. This means that, in order to extract the world population value from the HTML code, we will need to know something about how to perform **text processing**. We are going to need to search within the text to find the piece we want, and we are going to need to extract just that piece from the larger body of text.

The current population value on the web page is contained within the HTML code in a `div` tag that has an `id` attribute, with the unique

```
 1 <!DOCTYPE html PUBLIC "-//W3C//DTD XHTML 1.0 Transitional//EN"
 2     "http://www.w3.org/TR/xhtml1/DTD/xhtml1-transitional.dtd">
 3 <html xmlns="http://www.w3.org/1999/xhtml"
 4       xml:lang="en" lang="en">
 5 <head>
 6     <title>World POPClock Projection</title>
 7     <link rel="stylesheet"
 8           href="popclockworld%20Files/style.css"
 9           type="text/css">
10     <meta name="author" content="Population Division">
11     <meta http-equiv="Content-Type"
12           content="text/html; charset=iso-8859-1">
13     <meta name="keywords" content="world, population">
14     <meta name="description"
15           content="current world population estimate">
16     <style type="text/css">
17         #worldnumber {
18             text-align: center;
19             font-weight: bold;
20             font-size: 400%;
21             color: #ff0000;
22         }
23     </style>
24 </head>
25 <body>
26     <div id="cb_header">
27     <a href="http://www.census.gov/">
28     <img src="popclockworld%20Files/cb_head.gif"
29          alt="U.S. Census Bureau"
30          border="0" height="25" width="639">
31     </a>
32     </div>
33
34     <h1>World POPClock Projection</h1>
35
36     <p></p>
37     According to the <a href="http://www.census.gov/ipc/www/">
38     International Programs Center</a>, U.S. Census Bureau,
39     the total population of the World, projected to 09/12/07
40     at 07:05 GMT (EST+5) is<br><br>
41     <div id="worldnumber">6,617,746,521</div>
42     <p></p>
43     <hr>
       ...
```

Figure 9.3: Part of the HTML code for the World Population Clock web page (see Figure 9.2). The line numbers (in grey) are just for reference.

value "worldnumber" (line 41 in Figure 9.3). This makes it very easy to find the line that contains the population estimate because we just need to search for the pattern id="worldnumber". This text **search** task can be performed using the following code:

```
R> popLineNum <- grep('id="worldnumber"', clockHTML)
```

This code says "find the line of HTML code that contains the text id="worldnumber" and store the answer in RAM under the name popLineNum." The HTML code is fetched from RAM, we supply the pattern to search for by typing it at the keyboard, the computer searches the HTML code for our pattern and finds the matching line, and the result of our search is stored back in RAM.

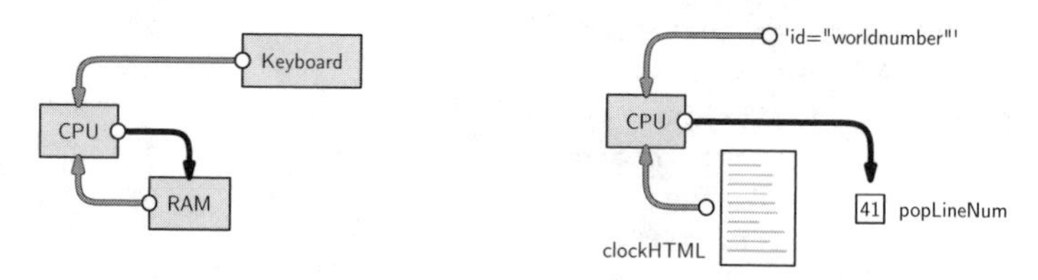

We can see the value that has been stored in RAM by typing the appropriate name.

```
R> popLineNum

[1] 41
```

Notice that the result this time is not text; it is a *number* representing the appropriate line within the HTML code.

Also notice that each time we store a value in RAM, we provide a label for the value so that we can access the value again later. We stored the complete set of HTML code with the label clockHTML, and we have now also stored the result of our search with the label popLineNum.

What we want is the actual line of HTML code rather than just the number telling us which line, so we need to use popLineNum to extract a **subset** of the text in clockHTML. This action is performed by the following code.

```
R> popLine <- clockHTML[popLineNum]
```

Again, this task involves using information that we already have in

RAM to calculate a new data value, and we store the new value back in RAM with the label popLine.

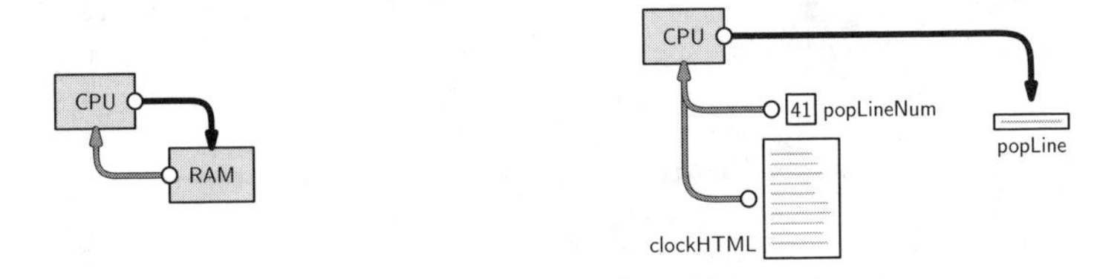

As before, we can just type the name to see the value that has been stored in RAM. The new value in this case is a line of text.

```
R> popLine

[1] "    <div id=\"worldnumber\">6,617,746,521</div>"
```

In many of the code examples throughout this chapter, we will follow this pattern: in one step, calculate a value and store it in RAM, with a label; then, in a second step, type the name of the label to display the value that has been stored.

Now that we have the important line of HTML code, we want to extract just the number, 6,617,746,521, from that line. This task consists of getting rid of the HTML tags. This is a text **search-and-replace** task and can be performed using the following code:

```
R> popText <- gsub('^.*<div id="worldnumber">|</div>.*$',
                   "", popLine)
R> popText

[1] "6,617,746,521"
```

This code says "delete the start and end div tags (and any spaces in front of the start tag)". We have used a **regular expression**, '^.*<div id="worldnumber">|</div>.*$', to specify the part of the text that we want to get rid of, and we have specified "", which means an empty piece of text, as the text to replace it with.

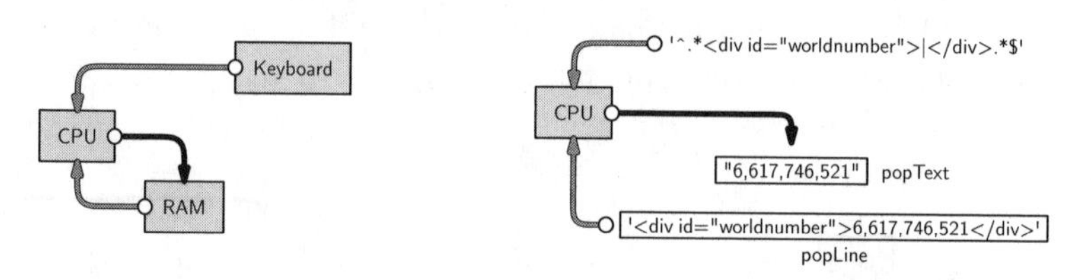

Section 9.9 describes text processing tasks and regular expressions in more detail.

At this point, we are close to having what we want, but we are not quite there yet because the value that we have for the world's population is still a piece of *text*, not a *number*. This is a very important point. We always need to be aware of exactly what sort of information we are dealing with. As described in Chapter 5, computers represent different sorts of values in different ways, and certain operations are only possible with certain types of data. For example, we ultimately want to be able to perform arithmetic on the population value that we are getting from this web site. That means that we must have a number; it does not make sense to perform arithmetic with text values.

Thus, the final thing we need to do is turn the text of the population estimate into a number so that we can later carry out mathematical operations. This process is called **type coercion** and appropriate code is shown below.

```
R> pop <- as.numeric(gsub(",", "", popText))
R> pop

[1] 6617746521
```

Notice that we have to process the text still further to remove the commas that are so useful for human viewers but a complete distraction for computers.

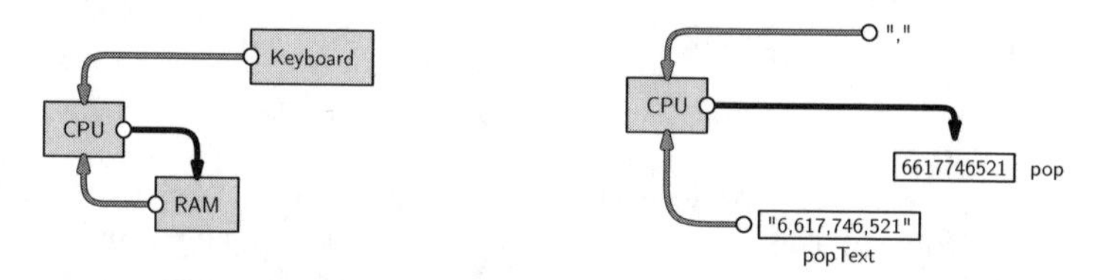

And now we have what we were after: the current U.S. Census Bureau estimate of the world's population from the World Population Clock

web site.

This first step provides a classic demonstration of the difference between performing a task by hand and writing code to get a computer to do the work. The manual method is simple, requires no new skills, and takes very little time. On the other hand, the computer code approach requires learning new information (it will take substantial chunks of this chapter to explain the code we have used so far), so it is more difficult and takes longer (the first time). However, the computer code approach *will* pay off in the long run, as we are about to see.

2. **Wait ten minutes.**

   The second step involves letting time pass so that we can obtain a second snapshot of the world population after a fixed time interval.

   Doing nothing is about as simple as it gets for a do-it-yourself task. However, it highlights two of the major advantages of automating tasks by computer. First, computers will perform boring tasks without complaining or falling asleep, and, second, their accuracy will not degrade as a function of the boredom of the task.

   The following code will make the computer wait for 10 minutes (600 seconds):

```
R> Sys.sleep(600)
```

3. **Copy the new value of the population clock.**

   The third step is to take another snapshot of the world population from the U.S. Census Bureau web site.

   This is the same as the first task. If we do it by hand, it is just as easy as it was before, though tasks like this quickly become tiresome if we have to repeat them many times.

   What about doing it by computer code?

   Here we see a third major benefit of writing computer code: once code has been written to perform a task, repetitions of the task become essentially free. All of the pain of writing the code in the first place starts to pay off very rapidly once a task has to be repeated. Almost exactly the same code as before will produce the new population clock estimate.

```
R> clockHTML2 <-
     readLines("http://www.census.gov/ipc/www/popclockworld.html")
```

```
R> popLineNum2 <- grep('id="worldnumber"', clockHTML2)
R> popLine2 <- clockHTML2[popLineNum2]
R> popText2 <- gsub('^.*<div id="worldnumber">|</div>.*$',
                    "", popLine2)
R> pop2 <- as.numeric(gsub(",", "", popText2))
R> pop2
```

```
[1] 6617747987
```

One detail that we have ignored to this point is the fact that the results of our calculations are being printed out. The information that we have stored in RAM is being displayed on the screen. As this example suggests, there may be differences between the value that is stored in memory and what is actually displayed for human eyes; in this case, the computer displays an "index", `[1]`, in front of the number. This is another important point to be aware of as we proceed through this chapter.

4. **Calculate the growth rate.**

   The fourth step in our task is to divide the change in the population estimate by the time interval.

   This is a very simple calculation that is, again, easy to do by hand. But arithmetic like this is just as easy to write code for. All we need to do is divide the change in population by the elapsed time (10 minutes):

```
R> rateEstimate <- (pop2 - pop)/10
R> rateEstimate
```

```
[1] 146.6
```

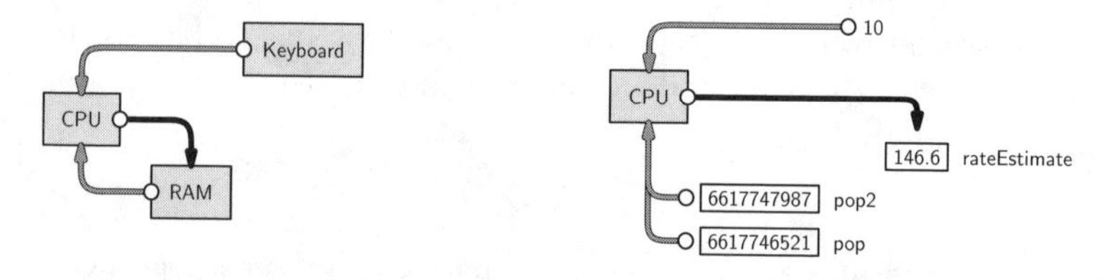

The final result is this: at the time of writing, we estimate that the world population was growing at the rate of about 147 people every minute.

As a final step, it would be prudent to save this result in a more permanent state, by writing this information to more permanent computer memory. The values that are in RAM will disappear when we quit from R. The following code creates a new text file and stores our rate estimate in that file.

```
R> writeLines(as.character(rateEstimate),
              "popRate.txt")
```

Notice that, in this step, we start with a number and convert it to a text value so that it can be stored as part of a text file.

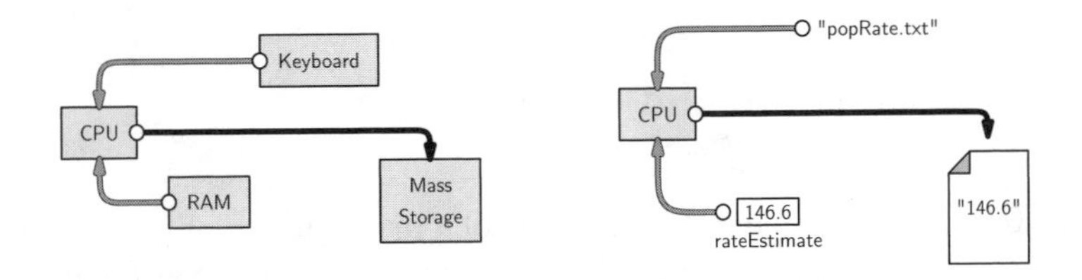

To reiterate, although that may seem like quite a lot of work to go through to perform a relatively simple task, the effort is worth it. By writing code so that the computer performs the task, we can improve our accuracy and efficiency, and we can repeat the task whenever we like for no additional cost. For example, we might want to improve our estimate of the population growth rate by taking several more snapshots from the population clock web site. This would take hours by hand, but we have most of the code already, and with a little more knowledge we could set the computer happily working away on this task for us.

This chapter is concerned with writing code like this, using the R language, to conduct and automate general data handling tasks: importing and exporting data, manipulating the shape of the data, and processing data into new forms.

In the following sections, we will begin to look specifically at how to perform these tasks in R.

9.2 The R environment

The name R is used to describe both the R language *and* the R software environment that is used to run code written in the language.

In this section, we will give a brief introduction to the R software. We will discuss the R language from Section 9.3 onwards.

The R software can be run on Windows, MacOS X, and Linux. An appropriate version may be downloaded from the Comprehensive R Archive Network (CRAN).[1] The user interface will vary between these settings, but the crucial common denominator that we need to know about is the **command line**.

Figure 9.4 shows what the command line looks like on Windows and on Linux.

9.2.1 The command line

The R command line interface consists of a **prompt**, usually the > character. We type code written in the R language and, when we press Enter, the code is run and the result is printed out. A very simple interaction with the command line looks like this:

```
R> 1 + 3 + 5 + 7

[1] 16
```

Throughout this chapter, examples of R code will displayed like this, with the R code preceded by a prompt, `R>`, and the results of the code (if any) displayed below the code. The format of the displayed result will vary because there can be many different kinds of results from running R code.

In this case, a simple arithmetic expression has been typed and the numeric result has been printed out.

[1] http://cran.r-project.org/

R Console

File Edit Misc Packages Help

```
R version 2.7.1 (2008-06-23)
Copyright (C) 2008 The R Foundation for Statistical Computing
ISBN 3-900051-07-0

R is free software and comes with ABSOLUTELY NO WARRANTY.
You are welcome to redistribute it under certain conditions.
Type 'license()' or 'licence()' for distribution details.

  Natural language support but running in an English locale

R is a collaborative project with many contributors.
Type 'contributors()' for more information and
'citation()' on how to cite R or R packages in publications.

Type 'demo()' for some demos, 'help()' for on-line help, or
'help.start()' for an HTML browser interface to help.
Type 'q()' to quit R.

> 1 + 3 + 5 + 7
[1] 16
> |
```

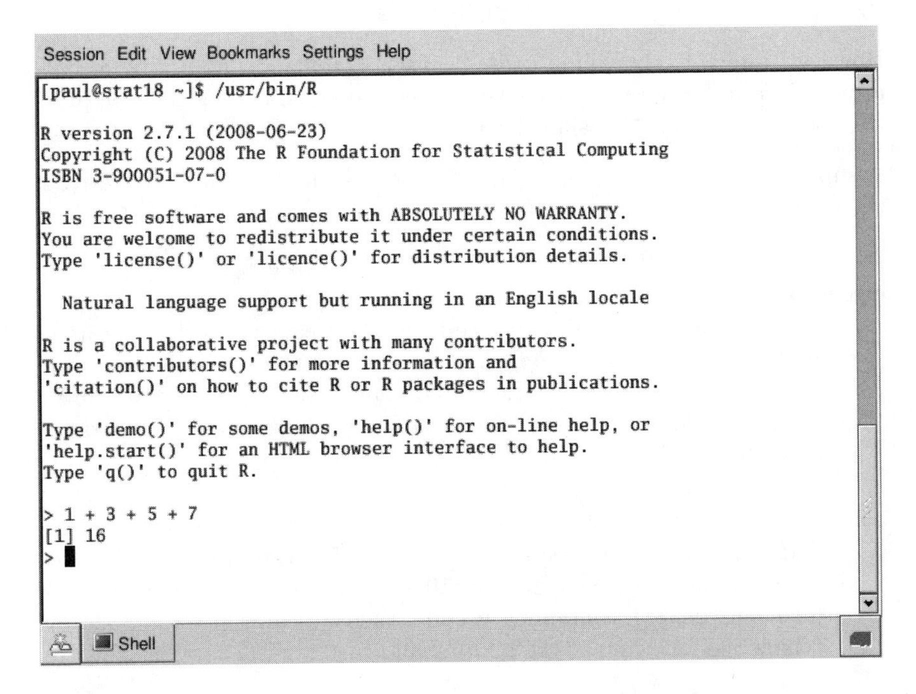

Figure 9.4: The R command line interface as it appears in the Windows GUI (top) and in an xterm on Linux (bottom).

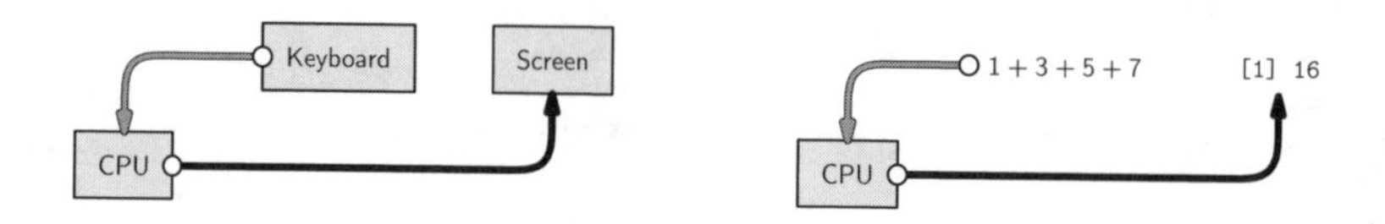

Notice that the result is not being stored in memory. We will look at how to retain results in memory in Section 9.3.

One way to write R code is simply to enter it interactively at the command line as shown above. This interactivity is beneficial for experimenting with R or for exploring a data set in a casual manner. For example, if we want to determine the result of division by zero in R, we can quickly find out by just trying it.

```
R> 1/0

[1] Inf
```

However, interactively typing code at the R command line is a very bad approach from the perspective of recording and documenting code because the code is lost when R is shut down.

A superior approach in general is to write R code in a file and get R to read the code from the file.

cut-and-paste

One way to work is to write R code in a text editor and then cut-and-paste bits of the code from the text editor into R. Some editors can be associated with an R session and allow submission of code chunks via a single key stroke (e.g., the Windows GUI provides a script editor with this facility).

`source()`

Another option is to read an entire file of R code into R using the `source()` function (see Section 10.3.8). For example, if we have a file called `code.R` containing R code, then we can run the R code by typing the following at the R command line:

```
R> source("code.R")
```

R reads the code from the file and runs it, one line at a time.

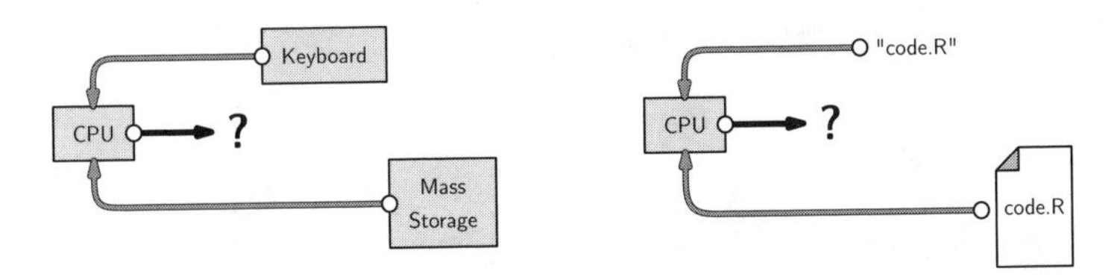

Whether there is any output and where it goes (to the screen, to RAM, or to mass storage) depends on the contents of the R code.

We will look at starting to write code using the R language in Section 9.3, but there is one example of R code that we need to know straight away. This is the code that allows us to exit from the R environment: to do this, we type `q()`.

9.2.2 The workspace

When quitting R, the option is given to save the "workspace image".

The workspace consists of all values that have been created during a session—all of the data values that have been stored in RAM.

The workspace is saved as a file called `.Rdata` and when R starts up, it checks for such a file in the current working directory and loads it automatically. This provides a simple way of retaining the results of calculations from one R session to the next.

However, saving the entire R workspace is not the recommended approach. It is better to save the original data set and R code and re-create results by running the code again.

If we have specific results that we want to save permanently to mass storage, for example, the final results of a large and time-consuming analysis, we can use the techniques described later in Sections 9.7 and 9.10.

The most important point for now is that we should save any *code* that we write; if we always know *how* we got a result, we can always recreate the result later, if necessary.

9.2.3 Packages

The features of R are organized into separate bundles called **packages**. The standard R installation includes about 25 of these packages, but many more can be downloaded from CRAN and installed to expand the things that R can do. For example, there is a package called **XML** that adds features for working with XML documents in R. We can **install** that package by typing the following code.

```
R> install.packages("XML")
```

Once a package has been installed, it must then be **loaded** within an R session to make the extra features available. For example, to make use of the **XML** package, we need to type the following code.

```
R> library("XML")
```

Of the 25 packages that are installed by default, nine packages are *loaded* by default when we start a new R session; these provide the basic functionality of R. All other packages must be loaded before the relevant features can be used.

Recap

The R environment is the software used to run R code.

R code is submitted to the R environment either by typing it directly at the command line, by cutting-and-pasting from a text file containing R code, or by specifying an entire file of R code to run.

R functionality is contained in packages. New functionality can be added by installing and then loading extra packages.

9.3 The R language

R is a popular programming language for working with and analyzing data.

As with the other computer languages that we have dealt with in this book, we have two main topics to cover for R: we need to learn the correct **syntax** for R code, so that we can write code that will run; and we need to learn the **semantics** of R code, so that we can make the computer do what we want it to do.

R is a programming language, which means that we can achieve a much greater variety of results with R compared to the other languages that we have seen. The cost of this flexibility is that the syntax for R is more complex and, because there are many more things that we can do, the R vocabulary that we have to learn is much larger.

In this section, we will look at the basics of R syntax—how to write correct R code. Each subsequent section will tackle the meaning of R code by focusing on a specific category of tasks and how they are performed using R.

9.3.1 Expressions

R code consists of one or more **expressions**.

An expression is an instruction to perform a particular task. For example, the following expression instructs R to add the first four odd numbers together.

```
R> 1 + 3 + 5 + 7

[1] 16
```

If there are several expressions, they are run, one at a time, in the order they appear.

The next few sections describe the basic types of R expressions.

9.3.2 Constant values

The simplest sort of R expression is just a constant value, typically a **numeric value** (a number) or a **character value** (a piece of text). For example, if we need to specify a number of seconds corresponding to 10 minutes, we specify a number.

```
R> 600
```

```
[1] 600
```

If we need to specify the name of a file that we want to read data from, we specify the name as a character value. Character values must be surrounded by either double-quotes or single-quotes.

```
R> "http://www.census.gov/ipc/www/popclockworld.html"
```

```
[1] "http://www.census.gov/ipc/www/popclockworld.html"
```

As shown above, the result of a constant expression is just the corresponding value and the result of an expression is usually printed out on the screen.

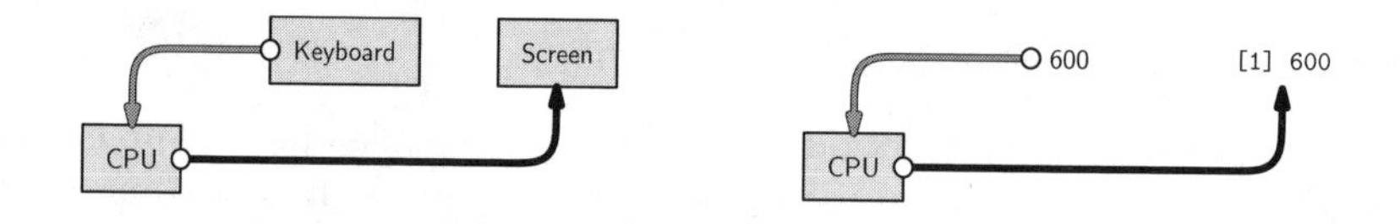

9.3.3 Arithmetic

An example of a slightly more complex expression is an arithmetic expression for calculating with numbers. R has the standard arithmetic **operators**:

`+`	addition.
`-`	subtraction.
`*`	multiplication.
`/`	division.
`^`	exponentiation.

For example, the following code shows the arithmetic calculation that was performed in Section 9.1 to obtain the rate of growth of the world's population—the change in population divided by the elapsed time. Note the use of parentheses to control the order of evaluation. R obeys the normal BODMAS rules of precedence for arithmetic operators, but parentheses are a useful way of avoiding any ambiguity, especially for a human audience.

```
R> (6617747987 - 6617746521) / 10

[1] 146.6
```

9.3.4 Conditions

A condition is an expression that has a yes/no answer—for example, whether one data value is greater than, less than, or equal to another. The result of a condition is a **logical value**: either `TRUE` or `FALSE`.

R has the standard operators for comparing values, plus operators for combining conditions:

`==`	equality.
`>` and `>=`	greater than (or equal to).
`<` and `<=`	less than (or equal to).
`!=`	inequality.
`&&`	logical and.
`||`	logical or.
`!`	logical not.

For example, the following code asks whether the second population estimate is larger than the first.

```
R> pop2 > pop

[1] TRUE
```

The code below asks whether the second population estimate is larger than the first *and* the first population estimate is greater than 6 billion.

```
R> (pop2 > pop) && (pop > 6000000000)

[1] TRUE
```

The parentheses in this code are not necessary, but they make the code easier to read.

9.3.5 Function calls

The most common and most useful type of R expression is a **function call**. Function calls are very important because they are how we use R to perform any non-trivial task.

A function call is essentially a complex instruction, and there are thousands of different R functions that perform different tasks. This section just looks at the basic *structure* of a function call; we will meet some important specific functions for data manipulation in later sections.

A function call consists of the function name followed by, within parentheses and separated from each other by commas, expressions called **arguments** that provide necessary information for the function to perform its task.

The following code gives an example of a function call that makes R pause for 10 minutes (600 seconds).

```
R> Sys.sleep(600)
```

The various components of this function call are shown below:

function name:	`Sys.sleep(600)`
parentheses:	`Sys.sleep(600)`
argument:	`Sys.sleep(600)`

The name of the function in this example is `Sys.sleep` (this function makes the computer wait, or "sleep", for a number of seconds). There is one argument to the function, the number of of seconds to wait, and in this case the value supplied for this argument is `600`.

Because function calls are so common and important, it is worth looking at a few more examples to show some of the variations in their format.

The `writeLines()` function saves text values into an external file. This function has two arguments: the text to save and the name of the file that the text will be saved into. The following expression shows a call to `writeLines()` that writes the text `"146.6"` into a file called `popRate.txt`.

```
R> writeLines("146.6", "popRate.txt")
```

The components of this function call are shown below:

function name:	`writeLines("146.6", "popRate.txt")`
parentheses:	`writeLines("146.6", "popRate.txt")`
comma between arguments:	`writeLines("146.6", "popRate.txt")`
first argument:	`writeLines("146.6", "popRate.txt")`
second argument:	`writeLines("146.6", "popRate.txt")`

This example demonstrates that commas must be placed between arguments in a function call. The first argument is the text to save, `"146.6"`, and the second argument is the name of the text file, `"popRate.txt"`.

The next example is very similar to the previous function call (the result is identical), but it demonstrates the fact that every argument has a name and these names can be specified as part of the function call.

```
R> writeLines(text="146.6", con="popRate.txt")
```

The important new components of this function call are shown below:

first arg. name:	`writeLines(text="146.6", con="popRate.txt")`
first arg. value:	`writeLines(text="146.6", con="popRate.txt")`
second arg. name:	`writeLines(text="146.6", con="popRate.txt")`
second arg. value:	`writeLines(text="146.6", con="popRate.txt")`

When arguments are named in a function call, they may be given in any order, so the previous function call would also work like this:

```
R> writeLines(con="popRate.txt", text="146.6")
```

The final example in this section is another variation on the function call to `writeLines()`.

```
R> writeLines("146.6")
```

```
146.6
```

The point of this example is that we can call the `writeLines()` function with only one argument. This demonstrates that some arguments have a **default value**, and if no value is supplied for the argument in the function call, then the default value is used. In this case, the default value of the `con` argument is a special value that means that the text is displayed on the screen rather than being saved to a file.

There are many other examples of function calls in the code examples throughout the remainder of this chapter.

9.3.6 Symbols and assignment

Anything that we type that starts with a letter, and which is not one of the special R keywords, is interpreted by R as a **symbol** (or **name**).

A symbol is a label for an object that is currently stored in RAM. When R encounters a symbol, it extracts from memory the value that has been stored with that label.

R automatically loads some predefined values into memory, with associated symbols. One example is the predefined symbol `pi`, which is a label for the the mathematical constant π.

```
R> pi
```

```
[1] 3.141593
```

The result of any expression can be **assigned** to a symbol, which means that the result is stored in RAM, with the symbol as its label.

For example, the following code performs an arithmetic calculation and stores the result in memory under the name `rateEstimate`.

```
R> rateEstimate <- (6617747987 - 6617746521) / 10
```

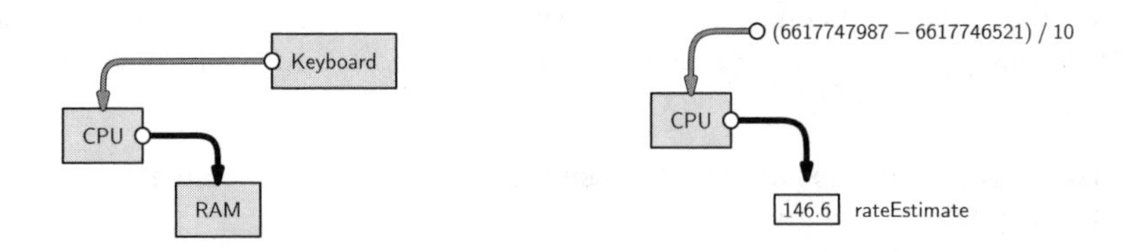

The combination of a less-than sign followed by a dash, `<-`, is called the **assignment operator**. We say that the symbol `rateEstimate` is assigned the value of the arithmetic expression.

Notice that R does not display any result after an assignment.

When we refer to the symbol `rateEstimate`, R will retrieve the corresponding value from RAM. An expression just consisting of a symbol will result in the value that is stored with that name being printed on the screen.

```
R> rateEstimate
```

```
[1] 146.6
```

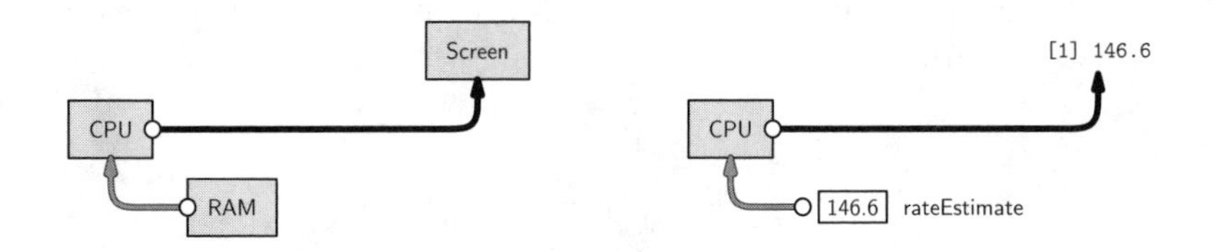

In many of the code examples throughout this chapter, we will follow this pattern: in one step, we will calculate a value and assign it to a symbol, which produces no display; then, in a second step, we will type the name of the symbol on its own to display the calculated value.

We can also use symbols as arguments to a function call, as in the following expression, which converts the numeric rate value into a text value.

```
R> as.character(rateEstimate)
```

```
[1] "146.6"
```

The value that has been assigned to `rateEstimate`, `146.6`, is retrieved from RAM and passed to the `as.character()` function, which produces a text version of the number.

In this case, we do not assign the result of our calculation, so it is automatically displayed on the screen.

For non-trivial tasks, assigning values to symbols is vital because we will need to perform several steps in order to achieve the ultimate goal. Assignments allow us to save intermediate steps along the way.

9.3.7 Keywords

Some symbols are used to represent special values. These predefined symbols cannot be re-assigned.

`NA`
: This symbol represents a missing or unknown value.

`Inf`
: This symbol is used to represent infinity (as the result of an arithmetic expression).

```
R> 1/0
[1] Inf
```

`NaN`
: This symbol is used to represent an arithmetic result that is undefined (Not A Number).

```
R> 0/0
[1] NaN
```

`NULL`
: This symbol is used to represent an empty result. Some R functions do not produce a result, so they return `NULL`.

`TRUE` **and** `FALSE`
: These symbols represent the logical values "true" and "false". The result of a condition is a logical value.

```
R> pop2 > pop
[1] TRUE
```

9.3.8 Flashback: Writing for an audience

Chapter 2 introduced general principles for writing computer code. In this section, we will look at some specific issues related to writing scripts in R.

The same principles, such as commenting code and laying out code so that it is easy for a human audience to read, still apply. In R, a comment is anything on a line after the special hash character, #. For example, the comment in the following line of code is useful as a reminder of why the number 600 has been chosen.

```
Sys.sleep(600) # Wait 10 minutes
```

Indenting is also very important. We need to consider indenting whenever an expression is too long and has to be broken across several lines of code. The example below shows a standard approach that left-aligns all arguments to a function call.

```
popText <- gsub('^.*<div id="worldnumber">|</div>.*$',
                "", popLine)
```

It is also important to make use of whitespace. Examples in the code above include the use of spaces around the assignment operator (<-), around arithmetic operators, and between arguments (after the comma) in function calls.

9.3.9 Naming variables

When writing R code, because we are constantly assigning intermediate values to symbols, we are forced to come up with lots of different symbol names. It is important that we choose sensible symbol names for several reasons:

1. Good symbol names are a form of documentation in themselves. A name like dateOfBirth tells the reader a lot more about what value has been assigned to the symbol than a name like d, or dob, or even date.

2. Short or convenient symbol names, such as x, or xx, or xxx should be avoided because it too easy to create errors by reusing the same name for two different purposes.

Anyone with children will know how difficult it can be to come up with even one good name, let alone a constant supply. However, unlike children, symbols usually have a specific purpose, so the symbol name naturally arises from a description of that purpose. A good symbol name should fully and accurately represent the information that has been assigned to that symbol.

One problem that arises is that a good symbol name is often a combination of two or more words. One approach to making such symbols readable is to use a mixture of lowercase and uppercase letters when typing the name; treat the symbol name like a sentence and start each new word with a capital letter (e.g., `dateOfBirth`). This naming mechanism is called "camelCase" (the uppercase letters form humps like the back of a camel).

Recap

A programming language is very flexible and powerful because it allows us to control the computer hardware as well as computer software.

R *is a programming language with good facilities for working with data.*

An instruction in the R *language is called an expression.*

Important types of expressions are: constant values, arithmetic expressions, function calls, and assignments.

Constant values are just numbers and pieces of text.

Arithmetic expressions are typed mostly as they would be written, except for division and exponentiation operators.

Function calls are instructions to perform a specific task and are of the form:

```
functionName(argument1, argument1)
```

An assignment saves a value in computer memory with an associated label, called a symbol, so that it can be retrieved again later. An assignment is of the form:

```
symbol <- expression
```

R *code should be written in a disciplined fashion just like any other computer code.*

9.4 Data types and data structures

We now have some knowledge of R syntax—what R expressions look like. Before we can start to learn some specific R expressions for particular data processing tasks, we first need to spend some time looking at how information is stored in computer memory.

When we are writing code in a programming language, we work most of the time with RAM, combining and restructuring data values to produce new values in RAM.

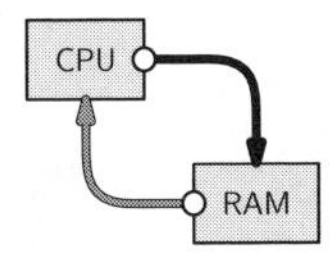

In Chapter 5, we looked at a number of different data storage formats.

In that discussion, we were dealing with long-term, persistent storage of information on mass storage computer memory.

Although, in this chapter, we will be working in RAM rather than with mass storage, we have exactly the same issues that we had in Chapter 5 of how to represent data values in computer memory. The computer memory in RAM is a series of 0's and 1's, just like the computer memory used to store files in mass storage. In order to work with data values, we need to get those values into RAM in some format.

At the basic level of representing a single number or a single piece of text, the solution is the same as it was in Chapter 5. Everything is represented as a pattern of bits, using various numbers of bytes for different sorts of values. In R, in an English locale, and on a 32-bit operating system, a single character usually takes up one byte, an integer takes four bytes, and a real number 8 bytes. Data values are stored in different ways depending on the **data type**—whether the values are numbers or text.

Although we do not often encounter the details of the memory representation, except when we need a rough estimate of how much RAM a data set might require, it is important to keep in mind what sort of data type we are working with because the computer code that we write will produce different results for different data types. For example, we can only calculate an average if we are dealing with values that have been stored as numbers, not if the values have been stored as text.

Another important issue is how *collections* of values are stored in memory.

The tasks that we will consider will typically involve working with an entire data set, or an entire variable from a data set, rather than just a single value, so we need to have a way to represent several related values in memory.

This is similar to the problem of deciding on a storage format for a data set, as we discussed in Chapter 5. However, rather than talking about different file formats, in this chapter we will talk about different **data structures** for storing a collection of data values in RAM. In this section, we will learn about the most common data structures in R.

Throughout this entire chapter, it will be important to always keep clear in our minds what data type we are working with and what sort of data structure are we working with.

Basic data types

Every individual data value has a data type that tells us what sort of value it is. The most common data types are numbers, which R calls **numeric values**, and text, which R calls **character values**.

We will meet some other data types as we progress through this section.

Basic data structures

We will look at five basic data structures that are available in R.

Because it is very important to know what sort of data structure we are dealing with, in the next few sections each result of an R expression will be accompanied by a small image (as shown below) that indicates the data structure involved.

Vectors

A collection of values that all have the same data type. The **elements** of a vector are all numbers, giving a **numeric vector**, or all character values, giving a **character vector**.

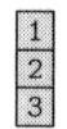

A vector can be used to represent a single variable in a data set.

Factors

A collection of values that all come from a fixed set of possible values. A factor is similar to a vector, except that the values within a factor are limited to a fixed set of possible values.

A factor can be used to represent a *categorical* variable in a data set.

Matrices

A two-dimensional collection of values that all have the same type. The values are arranged in rows and columns.

There is also an **array** data structure that extends this idea to more than two dimensions.

Data frames

A collection of vectors that all have the same length. This is like a matrix, except that each column can contain a different data type.

A data frame can be used to represent an entire data set.

Lists

A collection of data structures. The **components** of a list can be simply vectors—similar to a data frame, but with each column allowed to have a different length. However, a list can also be a much more complicated structure.

This is a very flexible data structure. Lists can be used to store any combination of data values together.

Starting with the next section, we will use a simple case study to explore the memory representation options available to us. We will also look at some of the functions that are used to create different data structures.

Table 9.1: The results from counting how many different sorts of candy there are in a bag of candy. There are 36 candies in total.

Shape	Pattern	Shade	Count
round	pattern	light	2
oval	pattern	light	0
long	pattern	light	3
round	plain	light	1
oval	plain	light	3
long	plain	light	2
round	pattern	dark	9
oval	pattern	dark	0
long	pattern	dark	2
round	plain	dark	1
oval	plain	dark	11
long	plain	dark	2

9.4.1 Case study: Counting candy

A counting puzzle. How many candies of each different shape are there? How many candies have a pattern? How many candies are dark and how many are light?

Table 9.1 shows the results of counting how many different sorts of candy there are in a bag of candy. The candies are categorized by their shape (round, oval, or long), their shade (light or dark), and whether they are plain or have a pattern.

In this example, we have information in a table that we can see (on a printed page or on a computer screen) and we want to enter this information into RAM by typing the values on the computer keyboard. We will look at how to write R code to store the information as data structures within RAM.

We will start by entering the first column of values from Table 9.1—the different shapes of candy. This will demonstrate the `c()` function for storing data as vectors.

```
R> shapes <- c("round", "oval", "long",
               "round", "oval", "long",
               "round", "oval", "long",
               "round", "oval", "long")
R> shapes
```

```
 [1] "round" "oval"  "long"  "round" "oval"  "long"  "round"
 [8] "oval"  "long"  "round" "oval"  "long"
```

The information in the first column consists of text, so each individual value is entered as a **character value** (within quotes), and the overall result is a **character vector**.

The result has been assigned to the symbol `shapes` so that we can use this character vector again later.

The second and third columns from Table 9.1 can be stored as character vectors in a similar manner.

```
R> patterns <- c("pattern", "pattern", "pattern",
                 "plain", "plain", "plain",
                 "pattern", "pattern", "pattern",
                 "plain", "plain", "plain")
R> patterns
```

```
 [1] "pattern" "pattern" "pattern" "plain"   "plain"
 [6] "plain"   "pattern" "pattern" "pattern" "plain"
[11] "plain"   "plain"
```

```
R> shades <- c("light", "light", "light",
               "light", "light", "light",
               "dark", "dark", "dark",
               "dark", "dark", "dark")
R> shades
```

```
 [1] "light" "light" "light" "light" "light" "light" "dark"
 [8] "dark"  "dark"  "dark"  "dark"  "dark"
```

The `c()` function also works with **numeric values**. In the following code, we create a **numeric vector** to store the fourth column of Table 9.1.

```
R> counts <- c(2, 0, 3, 1, 3, 2, 9, 0, 2, 1, 11, 2)
R> counts
```

```
 [1]  2  0  3  1  3  2  9  0  2  1 11  2
```

We now have the information from Table 9.1 stored as four vectors in RAM.

9.4.2 Vectors

The previous example demonstrated the `c()` function for concatenating values together to create a vector. In this section, we will look at some other functions that create vectors.

When data values have a regular pattern, the function `rep()` is extremely useful (`rep` is short for "repeat"). For example, column 1 of Table 9.1 has a simple pattern: the set of three possible shapes is repeated four times. The following code generates the `shapes` vector again, but this time using `rep()`.

```
R> shapes <- rep(c("round", "oval", "long"), 4)
R> shapes
```

```
 [1] "round" "oval"  "long"  "round" "oval"  "long"  "round"
 [8] "oval"  "long"  "round" "oval"  "long"
```

The first argument to `rep()` is the vector of values to repeat and the second argument says how many times to repeat the vector. The result is the original vector of 3 values repeated 4 times, producing a final vector with 12 values.

It becomes easier to keep track of what sort of data structure we are dealing with once we become familiar with the way that R displays the different types of data structures. With vectors, R displays an index inside square brackets at the start of each line of output, followed by the values, which are formatted so that they each take up the same amount of space (similar to a fixed-width file format). The previous result had room to display up to seven values on each row, so the second row of output starts with the eighth value (hence the `[8]` at the start of the line). All of the values have double-quotes around them to signal that these are all character values (i.e., this is a character vector).

As another example of the use of `rep()`, the following code generates a data structure containing the values in column 2 of Table 9.1.

```
R> patterns <- rep(c("pattern", "plain"), each=3, length=12)
R> patterns
```

```
 [1] "pattern" "pattern" "pattern" "plain"   "plain"
 [6] "plain"   "pattern" "pattern" "pattern" "plain"
[11] "plain"   "plain"
```

This example demonstrates two other arguments to `rep()`. The `each` argument says that each individual element of the original vector should be repeated 3 times. The `length` argument says that the final result should be a vector of length 12 (without that, the 2 original values would only be repeated 3 times each to produce a vector of 6 values; try it and see!).

To complete the set of variables in the candy data set, the following code generates the shade information from column 3 of Table 9.1.

```
R> shades <- rep(c("light", "dark"), each=6)
R> shades
```

```
 [1] "light" "light" "light" "light" "light" "light" "dark"
 [8] "dark"  "dark"  "dark"  "dark"  "dark"
```

The `rep()` function also works for generating numeric vectors; another important function for generating regular patterns of numeric values is the `seq()` function (`seq` is short for "sequence"). For example, a numeric vector containing the first 10 positive integers can be created with the following code.

```
R> seq(1, 10)
```

```
 [1]  1  2  3  4  5  6  7  8  9 10
```

The first argument to `seq()` specifies the number to start at and the second argument specifies the number to finish at.

There is also a `by` argument to allow steps in the sequence greater than one, as shown by the following code.

```
R> seq(1, 10, by=3)
```

```
[1]  1  4  7 10
```

For integer sequences in steps of 1, a short-hand equivalent is available using the special colon operator, :. For example, we could also have generated the first 10 positive integers with the code below.

```
R> 1:10
```

```
 [1]  1  2  3  4  5  6  7  8  9 10
```

Going back to the candy data, now that we have Table 9.1 stored as four vectors, we can begin to ask some questions about these data. This will allow us to show that when we perform a calculation with a vector of values, the result is often a new vector of values.

As an example, we will look at how to determine which types of candy did not appear in our bag of candy; in other words, we want to find the values in the counts vector that are equal to zero. The following code performs this calculation using a comparison.

```
R> counts == 0
```

```
 [1] FALSE  TRUE FALSE FALSE FALSE FALSE FALSE  TRUE FALSE
[10] FALSE FALSE FALSE
```

This result is interesting for three reasons. The first point is that the result of comparing two numbers for equality is a **logical value**: either TRUE or FALSE. Logical values are another of the basic data types in R. This result above is a **logical vector** containing only TRUE or FALSE values.

The second point is that we are starting with a numeric vector, counts, which contains 12 values, and we end up with a new vector that also has 12 values. In general, operations on vectors produce vectors as the result, so a very common way of generating new vectors is to do something with an existing vector.

The third point is that the 12 values in counts are being compared to a *single* numeric value, 0. The effect is to compare each of the 12 values separately against 0 and return 12 answers. This happens a lot in R when two vectors of different lengths are used. Section 9.6.1 discusses this idea of "recycling" shorter vectors further.

9.4.3 Factors

A **factor** is a basic data structure in R that is ideal for storing categorical data.

For example, consider the `shapes` vector that we created previously. This was just a character vector recording the text `"round"` for counts of round candies, `"oval"` for counts of oval candies, and `"long"` for counts of long candies.

This is not the ideal way to store this information because it does not acknowledge that elements containing the same text, e.g., `"round"`, really are the same value. A character vector can contain any text at all, so there are no data integrity constraints. The information would be represented better using a **factor**.

The following code creates the candy shape information as a factor:

```
R> shapesFactor <- factor(shapes,
                          levels=c("round", "oval", "long"))
R> shapesFactor

 [1] round oval  long  round oval  long  round oval  long
[10] round oval  long
Levels: round oval long
```

The first argument to the `factor()` function is the set of data values. We have also specified the set of valid values for the factor via the `levels` argument.

This is a better representation of the data because every value in the factor `shapesFactor` is now guaranteed to be one of the valid **levels** of the factor.

Factors are displayed in a similar way to vectors, but with additional information displayed about the levels of the factor.

9.4.4 Data frames

A vector in R contains values that are all of the same type. Vectors correspond to a single variable in a data set.

Most data sets consist of more than just one variable, so to store a complete data set we need a different data structure. In R, several variables can be stored together in an object called a **data frame**.

We will now build a data frame to contain all four variables in the candy data set (i.e., *all* of the information in Table 9.1).

The function `data.frame()` creates a data frame object from a set of vectors. For example, the following code generates a data frame from the variables that we have previously created, `shapes`, `patterns`, `shades`, and `counts`.

```
R> candy <- data.frame(shapes, patterns, shades, counts)
R> candy
```

```
   shapes patterns shades counts
1   round  pattern  light      2
2    oval  pattern  light      0
3    long  pattern  light      3
4   round    plain  light      1
5    oval    plain  light      3
6    long    plain  light      2
7   round  pattern   dark      9
8    oval  pattern   dark      0
9    long  pattern   dark      2
10  round    plain   dark      1
11   oval    plain   dark     11
12   long    plain   dark      2
```

We now have a data structure that contains the entire data set. This is a significant improvement over having four separate vectors because it properly represents the fact that the first value in each vector corresponds to information about the same type of candy.

An important feature of data frames is the fact that each column within a data frame can contain a different data type. For example, in the `candy` data frame, the first three columns contain text and the last column is numeric. However, all columns of a data frame must have the same length.

Data frames are displayed in a tabular layout, with column names above and row numbers to the left.

Another detail to notice about the way that data frames are displayed is that the text values in the first three columns do *not* have double-quotes around them (compare this with the display of text in character vectors in Section 9.4.2). Although character values must always be surrounded by double-quotes when we write code, they are not always *displayed* with double-quotes.

9.4.5 Lists

Vectors, factors, and data frames are the typical data structures that we create to represent our data values in computer memory. However, several other basic data structures are also important because when we call a function, the result could be any sort of data structure. We need to understand and be able to work with a variety of data structures.

As an example, consider the result of the following code:

```
R> dimnames(candy)
```

```
[[1]]
 [1] "1"  "2"  "3"  "4"  "5"  "6"  "7"  "8"  "9"  "10" "11"
[12] "12"

[[2]]
[1] "shapes"   "patterns" "shades"   "counts"
```

The `dimnames()` function extracts the column names and the row names from the `candy` data frame; these are the values that are displayed above and to the left of the data frame (see the example in the previous section). There are 4 columns and 12 rows, so the `dimnames()` function has to return two character vectors that have different lengths. A data frame can contain two vectors, but the vectors cannot have different lengths; the only way the `dimnames()` function can return these values is as a **list**.

In this case, we have a list with two components. The first component is a character vector containing the 12 row names and the second component is another character vector containing the 4 column names.

Notice the way that lists are displayed. The first component of the list starts with the component index, `[[1]]`, followed by the contents of this component, which is a character vector containing the names of the rows from the data frame.

```
[[1]]
 [1] "1"  "2"  "3"  "4"  "5"  "6"  "7"  "8"  "9"  "10" "11"
[12] "12"
```

The second component of the list starts with the component index [[2]], followed by the contents of this component, which is also a character vector, this time the column names.

```
[[2]]
[1] "shapes"   "patterns" "shades"   "counts"
```

The list() function can be used to create a list explicitly. Like the c() function, list() takes any number of arguments; if the arguments are named, those names are used for the components of the list.

In the following code, we generate a list similar to the previous one, containing the row and column names from the candy data frame.

```
R> list(rownames=rownames(candy),
        colnames=colnames(candy))

$rownames
 [1] "1"  "2"  "3"  "4"  "5"  "6"  "7"  "8"  "9"  "10" "11"
[12] "12"

$colnames
[1] "shapes"   "patterns" "shades"   "counts"
```

The difference is that, instead of calling dimnames() to get the entire list, we have called rownames() to get the row names as a character vector, colnames() to get the column names as another character vector, and then list() to combine the two character vectors into a list. The advantage is that we have been able to provide names for the components of the list. These names are evident in how the list is displayed on the screen.

A list is a very flexible data structure. It can have any number of **components**, each of which can be any data structure of any length or size. A simple example is a data-frame-like structure where each column can have a different length, but much more complex structures are also possible. For example, it is possible for a component of a list to be another list.

Anyone who has worked with a computer should be familiar with the idea of a list containing another list because a directory or folder of files has this sort of structure: a folder contains multiple files of different kinds and sizes and a folder can contain other folders, which can contain more files or even more folders, and so on. Lists allow for this kind of hierarchical structure.

9.4.6 Matrices and arrays

Another sort of data structure in R, which lies in between vectors and data frames, is the **matrix**. This is a two-dimensional structure (like a data frame), but one where all values are of the same type (like a vector).

As for lists, it is useful to know how to work with matrices because many R functions either return a matrix as their result or take a matrix as an argument.

A matrix can be created directly using the `matrix()` function. The following code creates a matrix from 6 values, with 3 columns and two rows; the values are used column-first.

```
R> matrix(1:6, ncol=3)
```

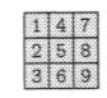

```
     [,1] [,2] [,3]
[1,]    1    3    5
[2,]    2    4    6
```

The **array** data structure extends the idea of a matrix to more than two dimensions. For example, a three-dimensional array corresponds to a data cube.

The `array()` function can be used to create an array. In the following code, a two-by-two-by-two, three-dimensional array is created.

```
R> array(1:8, dim=c(2, 2, 2))
```

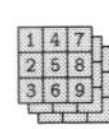

```
, , 1

     [,1] [,2]
[1,]    1    3
[2,]    2    4

, , 2

     [,1] [,2]
[1,]    5    7
[2,]    6    8
```

9.4.7 Flashback: Numbers in computer memory

Although we are not generally concerned with the bit-level or byte-level details of how data values are stored by R in RAM, we do need to be aware of one of the issues that was raised in Section 5.3.1.

In that section, we discussed the fact that there are limits to the precision with which numeric values can be represented in computer memory. This is true of numeric values stored in RAM just as it was for numeric values stored on some form of mass storage.

As a simple demonstration, consider the following condition.

```
R> 0.3 - 0.2 == 0.1
```

```
[1] FALSE
```

That apparently incorrect result is occurring because it is not actually possible to store an apparently simple value like 0.1 with absolute precision in computer memory (using a binary representation). The stored value is very, very close to 0.1, but it is not exact. In the condition above, the bit-level representations of the two values being compared are not identical, so the values, in computer memory, are not strictly equal.

Comparisons between real values must be performed with care and tests for equality between real values are not considered to be meaningful. The function `all.equal()` is useful for determining whether two real values are (approximately) equivalent.

Another issue is the precision to which numbers are *displayed*. Consider the following simple arithmetic expression.

```
R> 1/3
```

```
[1] 0.3333333
```

The answer, to full precision, never ends, but R has only shown seven significant digits. There is a limit to how many decimal places R *could* display because of the limits of representing numeric values in memory, but there is also a global **option** that controls (approximately) how many digits that R will display.

The following code uses the `options()` function to specify that R should display more significant digits.

```
R> options(digits=16)
R> 1/3
```

```
[1] 0.3333333333333333
```

The code below sets the option back to its default value.

```
R> options(digits=7)
```

Recap

A vector is a one-dimensional data structure and all of its elements are of the same data type.

A factor is one-dimensional and every element must be one of a fixed set of values, called the levels of the factor.

A matrix is a two-dimensional data structure and all of its elements are of the same type.

A data frame is two-dimensional and different columns may contain different data types, though all values within a column must be of the same data type and all columns must have the same length.

A list is a hierarchical data structure and each component of a list may be any type of data structure whatsoever.

9.5 Subsetting

Now that we know some basic R functions that allow us to enter data values and we have a basic idea of how data values are represented in RAM, we are in a position to start working with the data values.

One of the most basic ways that we can manipulate data structures is to **subset** them—select a smaller portion from a larger data structure. This is analogous to performing a query on a database.

For example, we might want to answer the question: "what sort of candy was the most common in the bag of candy?" The following code produces the answer to this question using R's subsetting capabilities.

```
R> candy[candy$counts == max(candy$counts), ]
```

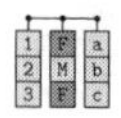

```
   shapes patterns shades counts
11   oval    plain   dark     11
```

R has very powerful mechanisms for subsetting. In this section, we will outline the basic format of these operations and many more examples will be demonstrated as we progress through the rest of the chapter.

We will start with subsetting vectors.

A subset from a vector may be obtained by appending an **index** within square brackets to the end of a symbol name. As an example, recall the vector of candy counts, called `counts`.

```
R> counts
```

```
 [1]  2  0  3  1  3  2  9  0  2  1 11  2
```

We can extract fourth of these counts by specifying the index `4`, as shown below.

```
R> counts[4]
```

```
[1] 1
```

The components of this expression are shown below:

symbol:	counts[4]
square brackets:	counts[4]
index:	counts[4]

The index can be a vector of any length. For example, the following code produces the first three counts (the number of light-shaded candies with a pattern).

```
R> counts[1:3]
```

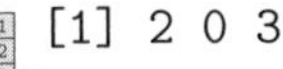

```
[1] 2 0 3
```

The diagram below illustrates how the values in the index vector are used to select elements from the `counts` vector to create a subset. Diagrams like this will be used throughout this chapter to help describe different data manipulation techniques; in each case, to save on space, the data values shown in the diagram will not necessarily correspond to the data values being discussed in the surrounding text.

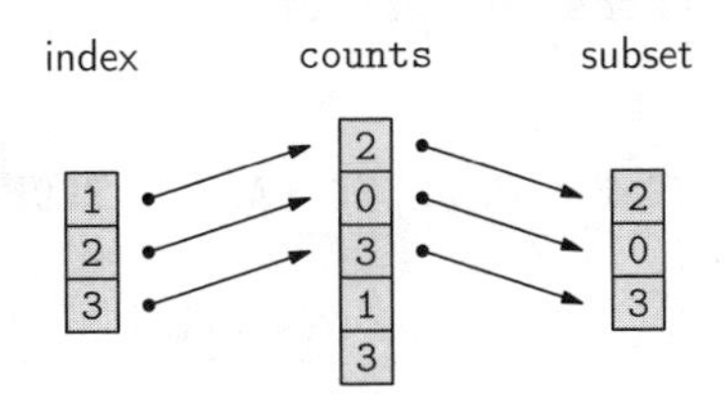

The index does not have to be a contiguous sequence, and it can include repetitions. The following example produces counts for all candies with a pattern. The elements we want are the first three *and* the seventh, eighth, and ninth. The index is constructed with the following code.

```
R> c(1:3, 7:9)
```

```
[1] 1 2 3 7 8 9
```

The subsetting operation is expressed as follows.

```
R> counts[c(1:3, 7:9)]
```

```
[1] 2 0 3 9 0 2
```

This is an example of a slightly more complex R expression. It involves a function call, to the `c()` function, that generates a vector, and this vector is then used as an index to select the corresponding elements from the `counts` vector.

The components of this expression are shown below:

symbol:	`counts[c(1:3, 7:9)]`
square brackets:	`counts[c(1:3, 7:9)]`
index:	`counts[c(1:3, 7:9)]`

As well as using integers for indices, we can use logical values. For example, a better way to express the idea that we want the counts for all candies with a pattern is to generate a logical vector first, as in the following code.

```
R> hasPattern <- patterns == "pattern"
R> hasPattern
```

```
 [1]  TRUE  TRUE  TRUE FALSE FALSE FALSE  TRUE  TRUE  TRUE
[10] FALSE FALSE FALSE
```

This logical vector can be used as an index to return all of the counts where `hasPattern` is `TRUE`.

```
R> counts[hasPattern]
```

```
[1] 2 0 3 9 0 2
```

The diagram below illustrates how an index of logical values selects elements from the complete object where the index value is `TRUE`.

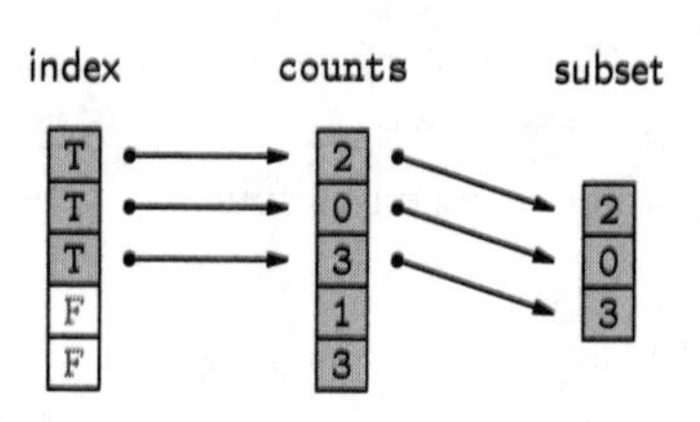

It would be even better to work with the entire data frame and retain the pattern with the counts, so that we can see that we have the correct result. We will now look at how subsetting works for two-dimensional data structures such as data frames.

A data frame can also be indexed using square brackets, though slightly differently because we have to specify both which rows *and* which columns we want. The following code extracts the **patterns** and **counts** variables, columns 2 and 4, from the data frame for all candies with a pattern:

```
R> candy[hasPattern, c(2, 4)]
```

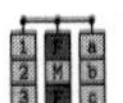

```
  patterns counts
1  pattern      2
2  pattern      0
3  pattern      3
7  pattern      9
8  pattern      0
9  pattern      2
```

The result is still a data frame, just a smaller one.

The components of this expression are shown below:

```
          symbol: candy[hasPattern, c(2, 4)]
 square brackets: count[hasPattern, c(2, 4)]
       row index: count[hasPattern, c(2, 4)]
           comma: count[hasPattern, c(2, 4)]
    column index: count[hasPattern, c(2, 4)]
```

An even better way to select this subset is to refer to the appropriate columns by their names. When a data structure has named components, a subset may be selected using those names. For example, the previous subset could also be obtained with the following code.

```
R> candy[hasPattern, c("patterns", "counts")]
```

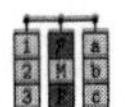

```
  patterns counts
1  pattern      2
2  pattern      0
3  pattern      3
7  pattern      9
8  pattern      0
9  pattern      2
```

The function **subset()** provides another way to subset a data frame. This function has a **subset** argument for specifying the rows and a **select** argument for specifying the columns.

```
R> subset(candy, subset=hasPattern,
         select=c("patterns", "counts"))
```

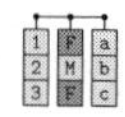

```
  patterns counts
1  pattern      2
2  pattern      0
3  pattern      3
7  pattern      9
8  pattern      0
9  pattern      2
```

When subsetting using square brackets, it is possible to leave the row or column index completely empty. The result is that all rows or all columns, respectively, are returned. For example, the following code extracts all columns for the first three rows of the data frame (the light-shaded candies with a pattern).

```
R> candy[1:3, ]
```

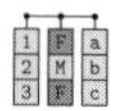

```
  shapes patterns shades counts
1  round  pattern  light      2
2   oval  pattern  light      0
3   long  pattern  light      3
```

If a single index is specified when subsetting a data frame with single square brackets, the effect is to extract the appropriate *columns* of the data frame and all rows are returned.

```
R> candy["counts"]
```

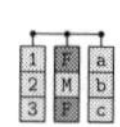

```
   counts
1       2
2       0
3       3
4       1
5       3
6       2
7       9
8       0
9       2
10      1
11     11
12      2
```

This result is one that we need to study more closely. This subsetting operation has extracted a single variable from a data frame. However, the result is a *data frame* containing a single column (i.e., a data set with one variable).

Often what we will require is just the *vector* representing the values in the variable. This is achieved using a different sort of indexing that uses *double* square brackets, `[[`. For example, the following code extracts the first variable from the `candy` data frame as a vector.

```
R> candy[["counts"]]
```

```
 [1]  2  0  3  1  3  2  9  0  2  1 11  2
```

The components of this expression are shown below:

symbol:	`candy[["counts"]]`
double square brackets:	`candy[["counts"]]`
index:	`candy[["counts"]]`

Single square bracket subsetting on a data frame is like taking an egg container that contains a dozen eggs and chopping up the *container* so that we are left with a smaller egg container that contains just a few eggs. Double square bracket subsetting on a data frame is like selecting just one *egg* from an egg container.

As with single square bracket subsetting, the index used for double square bracket subsetting can also be a number.

```
R> candy[[4]]
```

```
 [1]  2  0  3  1  3  2  9  0  2  1 11  2
```

However, with double square bracket subsetting, the index must be a single value.

There is also a short-hand equivalent for getting a single variable from a data frame. This involves appending a dollar sign, `$`, to the symbol, followed by the name of the variable.

```
R> candy$counts
```

```
 [1]  2  0  3  1  3  2  9  0  2  1 11  2
```

The components of this expression are shown below:

symbol:	`candy$counts` (candy underlined)
dollar sign:	`candy$counts` ($ underlined)
variable name:	`candy$counts` (counts underlined)

9.5.1 Assigning to a subset

The subsetting syntax can also be used to assign a new value to some portion of a larger data structure. As an example, we will look at replacing the zero values in the `counts` vector (the counts of candies) with a missing value, `NA`.

As with extracting subsets, the index can be a numeric vector, a character vector, or a logical vector. In this case, we will first develop an expression that generates a logical vector telling us where the zeroes are.

```
R> counts == 0
```

```
 [1] FALSE  TRUE FALSE FALSE FALSE FALSE FALSE  TRUE FALSE
[10] FALSE FALSE FALSE
```

The zeroes are the second and eighth values in the vector.

We can now use this expression as an index to specify which elements of the `counts` vector we want to modify.

```
R> counts[counts == 0] <- NA
R> counts
```

```
 [1]  2 NA  3  1  3  2  9 NA  2  1 11  2
```

We have replaced the original zero values with `NA`s.

The following code reverses the process and replaces all `NA` values with zero. The `is.na()` function is used to find which values within the `counts` vector are `NA`s.

```
R> counts[is.na(counts)] <- 0
```

9.5.2 Subsetting factors

The case of subsetting a factor deserves special mention because, when we subset a factor, the levels of the factor are not altered. For example, consider the `patterns` variable in the `candy` data set.

```
R> candy$patterns
```

```
 [1] pattern pattern pattern plain   plain   plain   pattern
 [8] pattern pattern plain   plain   plain
Levels: pattern plain
```

This factor has two levels, `pattern` and `plain`. If we subset just the first three values from this factor, the result only contains the value `pattern`, but there are still two levels.

```
R> candy$patterns[1:3]
```

```
[1] pattern pattern pattern
Levels: pattern plain
```

It is possible to force the unused levels to be dropped by specifying `drop=TRUE` within the square brackets, as shown below.

```
R> subPattern <- candy$patterns[1:3, drop=TRUE]
R> subPattern
```

```
[1] pattern pattern pattern
Levels: pattern
```

Assigning values to a subset of a factor is also a special case because only the current levels of the factor are allowed. A missing value is generated if the new value is not one of the current factor levels (and a warning is displayed).

For example, in the following code, we attempt to assign a new value to the first element of a factor where the new value is *not* one of the levels of the factor, so the result is an NA value.

```
R> subPattern[1] <- "swirly"
R> subPattern
```

```
[1] <NA>    pattern pattern
Levels: pattern
```

Recap

Single square brackets, [ ], select one or more elements from a data structure. The result is the same sort of data structure as the original object, just smaller.

The index within the square brackets can be a numeric vector, a logical vector, or a character vector.

Double square brackets, [[ ]], select a single element from a data structure. The result is usually a simpler data structure than the original.

The dollar sign, $, is short-hand for double square brackets.

9.6 More on data structures

This section provides some more details about how R data structures work. The information in this section is a little more advanced, but most of it is still useful for everyday use of R and most of it will be necessary to completely understand some of the R functions that are introduced in later sections.

9.6.1 The recycling rule

R allows us to work with vectors of values, rather than with single values, one at a time. This is very useful, but it does raise the issue of what to do when vectors have different lengths.

There is a general, but informal, rule in R that, in such cases, the shorter vector is recycled to become the same length as the longer vector. This is easiest to demonstrate via simple arithmetic.

In the following code, a vector of length 3 is added to a vector of length 6.

```
R> c(1, 2, 3) + c(1, 2, 3, 4, 5, 6)
```

```
[1] 2 4 6 5 7 9
```

What happens is that the first vector is recycled to make a vector of length 6, and then element-wise addition can occur.

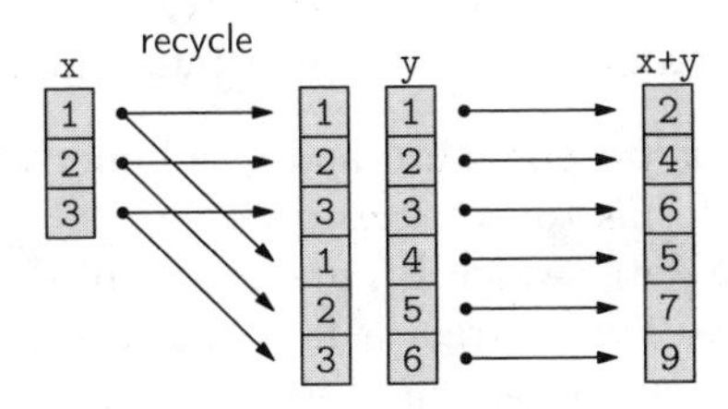

This rule is not necessarily followed in all possible situations, but it is the expected behavior in most cases.

9.6.2 Type coercion

In the case study in Section 9.1, there was a step where we took the text representation of a world population estimate and converted it into a number. This step is repeated below, broken down into a little more detail.

We start with a character vector (containing just one character value).

```
R> popText
```

```
[1] "6,617,746,521"
```

We remove the commas from the text, but we still have a character vector.

```
R> popNum <- gsub(",", "", popText)
R> popNum
```

```
[1] "6617746521"
```

Now, we convert the character vector to a numeric vector.

```
R> pop <- as.numeric(popNum)
R> pop
```

```
[1] 6617746521
```

The important part is the call to `as.numeric()`. This is the function that starts with a character value and converts it into a numeric value.

This process is called **type coercion** and it is important because we need the data in different forms to be able to perform different tasks. In this example, we need the data, which is an estimate of the world population, as a *number* so that we can subtract it from another, later, estimate of the world population. We cannot do the subtraction on character values.

There are many functions of the form `as.`*`type`*`()` for deliberately converting between different data structures like this. For example, the function for converting data into a character vector is called `as.character()`.

It is also important to keep in mind that many functions will *automatically* perform type coercion if we give them an argument in the wrong form. To demonstrate this, we will consider the `shapes` variable in the `candy` data frame.

The `shapes` vector that we created first is a *character vector*.

```
R> shapes
```

```
 [1] "round" "oval"  "long"  "round" "oval"  "long"  "round"
 [8] "oval"  "long"  "round" "oval"  "long"
```

We used this character vector, plus several others, to create the `candy` data frame, with a call to the `data.frame()` function.

```
R> candy <- data.frame(shapes, patterns, shades, counts)
```

What do we see if we extract the **shapes** column from the **candy** data frame?

```
R> candy$shapes

 [1] round oval  long  round oval  long  round oval  long
[10] round oval  long
Levels: long oval round
```

This is a *factor*, not a character vector!

How did the original character vector become a factor within the data frame? The `data.frame()` function automatically performed this type coercion (without telling us!).

This sort of automatic coercion happens a lot in R. Often, it is very convenient, but it is important to be aware that it may be happening, it is important to notice when it happens, and it is important to know how to stop it from happening. In some cases it will be more appropriate to perform type coercion explicitly, with functions such as `as.numeric()` and `as.character()`, and in some cases functions will provide arguments that turn the coercion off. For example, the `data.frame()` function provides a `stringsAsFactors` argument that controls whether character data are automatically converted to a factor.

Coercing factors

As we saw with subsetting factors, performing type coercion requires special care when we are coercing *from* a factor to another data type.

The correct sequence for coercing a factor is to first coerce it to a character vector and then to coerce the character vector to something else.

In particular, when coercing a factor that has levels consisting entirely of digits, the temptation is to call `as.numeric()` directly. However, the correct approach is to call `as.character()` and then `as.numeric()`.

9.6.3 Attributes

Consider the candy data frame again.

```
R> candy
```

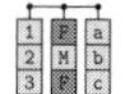

```
   shapes patterns shades counts
1   round  pattern  light      2
2    oval  pattern  light      0
3    long  pattern  light      3
4   round    plain  light      1
5    oval    plain  light      3
6    long    plain  light      2
7   round  pattern   dark      9
8    oval  pattern   dark      0
9    long  pattern   dark      2
10  round    plain   dark      1
11   oval    plain   dark     11
12   long    plain   dark      2
```

This data set consists of four variables that record how many candies were counted for each combination of shape, pattern, and shade. These variables are stored in the data frame as three factors and a numeric vector.

The data set also consists of some additional metadata. For example, each variable has a name. How is this information stored as part of the data frame data structure?

The answer is that the column names are stored as **attributes** of the data frame.

Any data structure in R may have additional information attached to it as an attribute; attributes are like an extra list structure attached to the primary data structure. The diagram below illustrates the idea of a data frame, on the left, with a list of attributes containing row names and column names, on the right.

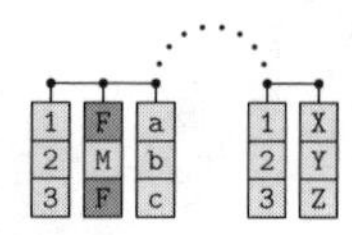

In this case, the data frame has an attribute containing the row names and another attribute containing the column names.

As another example, consider the factor `shapesFactor` that we created on page 237.

```
R> shapesFactor
```

```
 [1] round oval  long  round oval  long  round oval  long
[10] round oval  long
Levels: round oval long
```

Again, there are the actual data values, *plus* there is metadata that records the set of valid data values—the levels of the factor. This levels information is stored in an attribute. The diagram below illustrates the idea of a factor, on the left, with a single attribute that contains the valid levels of the factor, on the right.

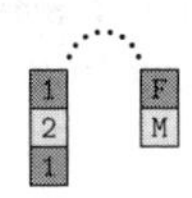

We do not usually work directly with the attributes of factors or data frames. However, some R functions produce a result that has important information stored in the attributes of a data structure, so it is necessary to be able to work with the attributes of a data structure.

The `attributes()` function can be used to view all attributes of a data structure. For example, the following code extracts the attributes from the `candy` data frame (we will talk about the `class` attribute later).

```
R> attributes(candy)
```

```
$names
[1] "shapes"   "patterns" "shades"   "counts"

$row.names
 [1]  1  2  3  4  5  6  7  8  9 10 11 12

$class
[1] "data.frame"
```

This is another result that we should look at very closely. What sort of data structure is it?

This is a *list* of attributes.

Each component of this list, each attribute of the data frame, has a name; the column names of the **candy** data frame are in a component of this attribute list called **names** and the row names of the **candy** data frame are in a component of this list called **row.names**.

The **attr()** function can be used to get just one specific attribute from a data structure. For example, the following code just gets the **names** attribute from the **candy** data frame.

```
R> attr(candy, "names")
```

```
[1] "shapes"   "patterns" "shades"   "counts"
```

Because many different data structures have a **names** attribute, there is a special function, **names()**, for extracting just that attribute.

```
R> names(candy)
```

```
[1] "shapes"   "patterns" "shades"   "counts"
```

For the case of a data frame, we could also use the **colnames()** function to get this attribute, and there is a **rownames()** function for obtaining the row names.

Similarly, there is a special function, **levels()**, for obtaining the levels of a factor.

```
R> levels(shapesFactor)
```

```
[1] "round" "oval"  "long"
```

Section 9.9.3 contains an example where it is necessary to directly access the attributes of a data structure.

9.6.4 Classes

We can often get some idea of what sort of data structure we are working with by simply viewing how the data values are displayed on screen. However, a more definitive answer can be obtained by calling the `class()` function.

For example, the data structure that has been assigned to the symbol `candy` is a data frame.

```
R> class(candy)

[1] "data.frame"
```

The `shapes` variable within the `candy` data frame is a factor.

```
R> class(candy$shapes)

[1] "factor"
```

Many R functions return a data structure that is not one of the basic data structures that we have already seen. For example, consider the following code, which generates a table of counts of the number of candies of each different shape (summing across shades and patterns).

We will describe the `xtabs()` function later in Section 9.8.4. For now, we are just interested in the data structure that is produced by this function.

```
R> shapeCount <- xtabs(counts ~ shapes, data=candy)
R> shapeCount

shapes
 long  oval round
    9    14    13
```

What sort of data structure is this? The best way to find out is to use the `class()` function.

```
R> class(shapeCount)
```

```
[1] "xtabs" "table"
```

The result is an "xtabs" data structure, which is a special sort of "table" data structure.

We have not seen either of these data structures before. However, much of what we already know about working with the standard data structures, and some of what we will learn in later sections, will also work with any new class that we encounter.

For example, it is usually possible to subset any class using the standard square bracket syntax. For example, in the following code, we extract the first element from the table.

```
R> shapeCount[1]
```

```
long
   9
```

Where appropriate, arithmetic and comparisons will also generally work. In the code below, we are calculating which elements of the table are greater than 10.

```
R> shapeCount > 10
```

```
shapes
 long  oval round
FALSE  TRUE  TRUE
```

Furthermore, if necessary, we can often resort to coercing a class to something more standard and familiar. The following code converts the table data structure into a data frame, where the rows of the original table have been converted into columns in the data frame, with appropriate column names automatically provided.

```
R> as.data.frame(shapeCount)
```

```
  shapes Freq
1   long    9
2   oval   14
3  round   13
```

In summary, although we will encounter a wide variety of data structures, the standard techniques that we learn in this chapter for working with basic data structures, such as subsetting, will also be useful for working with other classes.

9.6.5 Dates

Dates are an important example of a special data structure. Representing dates as just text is convenient for humans to view, but other representations are better for computers to work with.

As an example, we will make use of the `Sys.Date()` function, which returns the current date.

```
R> today <- Sys.Date()
R> today
```

```
[1] "2008-09-29"
```

This looks like it is a character vector, but it is not. It is a **Date** data structure.

```
R> class(today)
```

```
[1] "Date"
```

Having a special class for dates means that we can perform tasks with dates, such as arithmetic and comparisons, in a meaningful way, something we could not do if we stored the date as just a character value. For example, the manuscript for this book was due at the publisher on September 30th 2008. The following code calculates whether the manuscript was late. The `as.Date()` function converts a character value, in this case `"2008-09-30"`, into a date.

```
R> deadline <- as.Date("2008-09-30")
R> today > deadline
```

```
[1] FALSE
```

The following code calculates how many days remain before the deadline (or how late the manuscript is).

```
R> deadline - today
```

```
Time difference of 1 days
```

The Date class stores date values as integer values, representing the number of days since January 1st 1970, and automatically converts the numbers to a readable text value to display the dates on the screen.

9.6.6 Formulas

Another important example of a special class in R is the **formula** class.

A formula is created by the special tilde operator, `~`. Formulas are created to describe relationships using symbols.

We saw an example on page 259 that looked like this:

```
R> counts ~ shapes
```

```
counts ~ shapes
```

The components of this expression are show below:

left-hand side:	<u>`counts`</u> `~ shapes`
tilde operator:	`counts` <u>`~`</u> `shapes`
right-hand side:	`counts ~` <u>`shapes`</u>

The result of a formula is just the formula expression itself. A formula only involves the symbol names; any data values that have been assigned to the symbols in the formula are not accessed at all in the creation of the formula.

Each side of a formula can have more than one symbol, with the symbols separated by standard operators such as `+` and `*`.

For example, the following call to the `xtabs()` function combines two symbols on the right-hand side of the formula to produce a two-way table of counts.

```
R> xtabs(counts ~ patterns + shades, data=candy)
```

```
xtabs(counts ~ patterns + shades, data=candy)
```

Formulas are mainly used to express statistical models in R, but they are also used in other contexts, such as the `xtabs()` function shown above and in Section 9.6.4. We will see another use of formulas in Section 9.8.11.

9.6.7 Exploring objects

When working with anything but tiny data sets, basic features of the data set cannot be determined by just viewing the data values. This section describes a number of functions that are useful for obtaining useful summary features from a data structure.

The `summary()` function produces summary information for a data structure. For example, it will provide numerical summary data for each variable in a data frame.

```
R> summary(candy)
```

```
   shapes      patterns     shades       counts
 long :4   pattern:6   dark :6   Min.   : 0
 oval :4   plain  :6   light:6   1st Qu.: 1
 round:4                         Median : 2
                                 Mean   : 3
                                 3rd Qu.: 3
                                 Max.   :11
```

The `length()` function is useful for determining the number of values in a vector or the number of components in a list. Similarly, the `dim()` function will give the number of rows and columns in a matrix or data frame.

```
R> dim(candy)
```

```
[1] 12  4
```

The `str()` function (short for "structure") is useful when dealing with large objects because it only shows a sample of the values in each part of the object, although the display is very low-level so it may not always make things clearer.

The following code displays the low-level structure of the `candy` data frame.

```
R> str(candy)
```

```
'data.frame':        12 obs. of  4 variables:
 $ shapes  : Factor w/ 3 levels "long","oval",..: 3 2 1 3 ..
 $ patterns: Factor w/ 2 levels "pattern","plain": 1 1 1 2..
 $ shades  : Factor w/ 2 levels "dark","light": 2 2 2 2 2 ..
 $ counts  : num  2 0 3 1 3 2 9 0 2 1 ...
```

Another function that is useful for inspecting a large object is the `head()` function. This just shows the first few elements of an object, so we can see the basic structure without seeing all of the values. The code below uses `head()` to display only the first six rows of the `candy` data frame.

```
R> head(candy)
```

```
  shapes patterns shades counts
1  round  pattern  light      2
2   oval  pattern  light      0
3   long  pattern  light      3
4  round    plain  light      1
5   oval    plain  light      3
6   long    plain  light      2
```

There is also a `tail()` function for viewing the last few elements of an object.

9.6.8 Generic functions

In Section 9.6.2, we saw that some functions automatically perform type coercion. An example is the `paste()` function, which combines text values. If we give it a value that is not text, it will automatically coerce it to text. For example, the following code returns a number (the total number of candies in the `candy` data frame).

```
R> sum(candy$counts)

[1] 36
```

If we use this value as an argument to the paste() function, the number is automatically coerced to a text value to become part of the overall text result.

```
R> paste("There are", sum(candy$counts), "long candies")

[1] "There are 36 long candies"
```

Generic functions are similar in that they will accept many different data structures as arguments. However, instead of forcing the argument to be what the function wants it to be, a generic function adapts itself to the data structure it is given. Generic functions do different things when given different data structures.

An example of a generic function is the summary() function. The result of a call to summary() will depend on what sort of data structure we provide. The summary information for a factor is a table giving how many times each level of the factor appears.

```
R> summary(candy$shapes)

 long  oval round 
    4     4     4 
```

If we provide a numeric vector, the result is a five-number summary, plus the mean.

```
R> summary(candy$counts)

   Min. 1st Qu.  Median    Mean 3rd Qu.    Max. 
      0       1       2       3       3      11 
```

Generic functions are another reason why it is easy to work with data in R; a single function will produce a sensible result no matter what data structure we provide.

However, generic functions are also another reason why it is so important to be aware of what data structure we are working with. Without knowing what sort of data we are using, we cannot know what sort of result to expect from a generic function.

Recap

When two vectors of different lengths are used together, the shorter vector is often recycled to make it the same length as the longer vector.

Type coercion is the conversion of data values from one data type or data structure to another. This may happen automatically within a function, so care should be taken to make sure that data values returned by a function have the expected data type or data structure.

Any data structure may have attributes, which provide additional information about the data structure. These are in the form of a list of data values that are attached to the main data structure.

All data structures have a class.

Basic data manipulations such as subsetting, arithmetic, and comparisons, should still work even with unfamiliar classes.

Dates and formulas are important examples of special classes.

A generic function is a function that produces a different result depending on the class of its arguments.

9.7 Data import/export

Almost all of the examples so far have used data that are typed explicitly as R expressions. In practice, data usually reside in one or more files of various formats and in some form of mass storage. This section looks at R functions that can be used to read data into R from such external files.

We will look at functions that deal with all of the different data storage options that were discussed in Chapter 5: plain text files, XML documents, binary files, spreadsheets, and databases.

We will also look at some functions that go the other way and write a data structure from RAM to external mass storage.

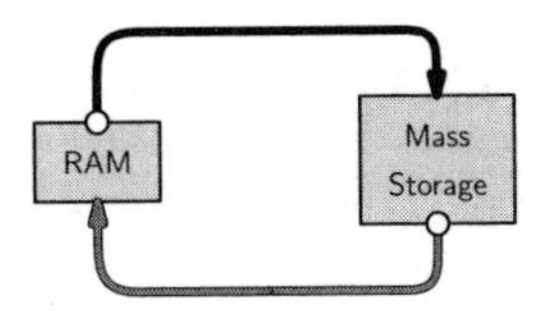

9.7.1 The working directory

This section provides a little more information about how the R software environment works, with respect to reading and writing files.

Any files created during an R session are created in the current **working directory** of the session, unless an explicit path or folder or directory is specified. Similarly, when files are read into an R session, they are read from the current working directory.

On Linux, the working directory is the directory that the R session was started in. This means that the standard way to work on Linux is to create a directory for a particular project, put any relevant files in that directory, change into that directory, and then start an R session.

On Windows, it is typical to start R by double-clicking a shortcut or by selecting from the list of programs in the 'Start' menu. This approach will, by default, set the working directory to one of the directories where R was installed, which is a bad place to work. Instead, it is a good idea to work in a separate directory for each project, create a shortcut to R within that directory, and set the 'Start in' field on the properties dialog of the shortcut to be the directory for that project. An alternative is to use the `setwd()` function or the 'Change dir' option on the 'File' menu to explicitly change the working directory to something appropriate when working on a particular project.

9.7.2 Specifying files

In order to read or write a file, the first thing we need to be able to do is specify *which* file we want to work with. Any function that works with a file requires a precise description of the name of the file and the location of the file.

A filename is just a character value, e.g., "pointnemotemp.txt", but specifying the location of a file can involve a **path**, which describes a location on a persistent storage medium, such as a hard drive.

The best way to specify a path in R is via the file.path() function because this avoids the differences between path descriptions on different operating systems. For example, the following code generates a path to the file pointnemotemp.txt within the directory LAS (on a Linux system).

```
R> file.path("LAS", "pointnemotemp.txt")

[1] "LAS/pointnemotemp.txt"
```

The file.choose() function can be used to allow interactive selection of a file. This is particularly effective on Windows because it provides a familiar file selection dialog box.

9.7.3 Text formats

R has functions for reading in the standard types of plain text formats: delimited formats, especially CSV files, and fixed-width formats (see Section 5.2). We will briefly describe the most important functions and then demonstrate their use in an example.

The read.table() function works for data in a delimited format. By default, the delimiter is whitespace (spaces and tabs), but an alternative may be specified via the sep argument. There is a read.csv() function for the special case of CSV files, and for data in a fixed-width format, there is the read.fwf() function.

The important arguments to these functions are the name of the external file and information about the format of the file. For example, in order to read a file in a fixed-width format with the read.fwf() function, we have to supply the widths of the fields in the file via the widths argument.

The result returned by all of these functions is a *data frame.*

Another important piece of information required when reading text files is the data type of each column of data in the file. Everything in the file is text, including numeric values, which are stored as a series of digits, and this means that some of the text values from the file may need to be coerced so that they are stored as an appropriate data type in the resulting data frame.

The general rule is that, if all values within a column in the text file are numbers, then the values in that column are coerced to a numeric vector. Otherwise, the values are used to create a *factor*. Several arguments are provided to control the coercion from the text values in the file to a specific data type; we will see examples in the next section and in other case studies throughout this chapter.

Another function that can be used to read text files is the `readLines()` function. The result in this case is a character vector, where each line of the text file becomes a separate element of the vector. The text values from the file become text values within RAM, so no type coercion is necessary in this case.

This function is useful for processing a file that contains text, but not in a standard plain text format. For example, the `readLines()` function was used to read the HTML code from the World Population Clock web site in Section 9.1. Section 9.9 discusses tools for processing data that have been read into R as text.

The following case study provides some demonstrations of the use of these functions for reading text files.

9.7.4 Case study: Point Nemo (continued)

The temperature data obtained from NASA's Live Access Server for the Pacific Pole of Inaccessibility (see Section 1.1) were delivered in a plain text format (see Figure 9.5, which reproduces Figure 1.2 for convenience). How can we load this temperature information into R?

A `read.table()` example

One way to view the format of the file in Figure 9.5 is that the data start on the ninth line and data values within each row are separated by whitespace. This means that we can use the `read.table()` function to read the Point Nemo temperature information, as shown below.

```
R> pointnemodelim <-
       read.table(file.path("LAS", "pointnemotemp.txt"),
                  skip=8)
```

```
            VARIABLE : Mean TS from clear sky composite (kelvin)
            FILENAME : ISCCPMonthly_avg.nc
            FILEPATH : /usr/local/fer_data/data/
            SUBSET   : 48 points (TIME)
            LONGITUDE: 123.8W(-123.8)
            LATITUDE : 48.8S
                      123.8W
                       23
 16-JAN-1994 00 /  1:  278.9
 16-FEB-1994 00 /  2:  280.0
 16-MAR-1994 00 /  3:  278.9
 16-APR-1994 00 /  4:  278.9
 16-MAY-1994 00 /  5:  277.8
 16-JUN-1994 00 /  6:  276.1
 ...
```

Figure 9.5: The first few lines of the plain text output from the Live Access Server for the surface temperature at Point Nemo. This is a reproduction of Figure 1.2.

```
R> pointnemodelim
```

```
            V1 V2 V3 V4    V5
1 16-JAN-1994  0  / 1: 278.9
2 16-FEB-1994  0  / 2: 280.0
3 16-MAR-1994  0  / 3: 278.9
4 16-APR-1994  0  / 4: 278.9
5 16-MAY-1994  0  / 5: 277.8
6 16-JUN-1994  0  / 6: 276.1
...
```

In the above example, and in a number of examples throughout the rest of the chapter, the output displayed by R has been manually truncated to avoid wasting too much space on the page. This truncation is indicated by the use of ... at the end of a line of output, to indicate that there are further columns that are not shown, or ... on a line by itself, to indicate that there are further rows that are not shown.

By default, `read.table()` assumes that the text file contains a data set with one case on each row and that each row contains multiple values, with each value separated by whitespace (one or more spaces or tabs). The `skip` argument is used to ignore the first few lines of a file when, for example, there is header information or metadata at the start of the file before the core data values.

The result of this function call is a data frame, with a variable for each column of values in the text file. In this case, there are four instances of whitespace on each line, so each line is split into five separate values, resulting in a data frame with five columns.

The types of variables are determined automatically. If a column only contains numbers, the variable is numeric; otherwise, the variable is a factor.

The names of the variables in the data frame can be read from the file, or specified explicitly in the call to `read.table()`. Otherwise, as in this case, R will generate a unique name for each column: `V1`, `V2`, `V3`, etc.

The result in this case is not perfect because we end up with several columns of junk that we do not want (`V2` to `V4`). We can use a few more arguments to `read.table()` to improve things greatly.

```
R> pointnemodelim <-
       read.table(file.path("LAS", "pointnemotemp.txt"),
                  skip=8,
                  colClasses=c("character",
                               "NULL", "NULL", "NULL",
                               "numeric"),
                  col.names=c("date", "", "", "", "temp"))
```

```
R> pointnemodelim
```

```
         date  temp
1 16-JAN-1994 278.9
2 16-FEB-1994 280.0
3 16-MAR-1994 278.9
4 16-APR-1994 278.9
5 16-MAY-1994 277.8
6 16-JUN-1994 276.1
...
```

The `colClasses` argument allows us to control the types of the variables explicitly. In this case, we have forced the first variable to be just text (these values are dates, not categories). There are five columns of values in the text file (treating whitespace as a column break), but we are not interested in the middle three, so we use "NULL" to indicate that these columns should just be ignored. The last column, the temperature values, is numeric.

It is common for the names of the variables to be included as the first line of a text file (the header argument can be used to read variable names from such a file). In this case, there is no line of column names in the file, so we provide the variable names explicitly, as a character vector, using the col.names argument.

The dates can be converted from character values to date values in a separate step using the as.Date() function.

```
R> pointnemodelim$date <- as.Date(pointnemodelim$date,
                                  format="%d-%b-%Y")
```

```
R> pointnemodelim
```

```
        date  temp
1 1994-01-16 278.9
2 1994-02-16 280.0
3 1994-03-16 278.9
4 1994-04-16 278.9
5 1994-05-16 277.8
6 1994-06-16 276.1
...
```

The format argument contains special sequences that tell as.Date() where the various components of the date are within the character values. The %d means that there are two digits for the day, the %b means that the month is given as an abbreviated month name, and the %Y means that there are four digits for the year. The dashes are literal dash characters.

The way that these components map to the original character value for the first date is shown below.

two-digit day, %d:	16-JAN-1994
abbreviated month name, %b:	16-JAN-1994
four-digit year, %Y:	16-JAN-1994
literal dashes:	16-JAN-1994

Thus, for example, the original *character* value 16-JAN-1994 becomes the *date* value 1994-01-16.

A `read.fwf()` example

Another way to view the Point Nemo text file in Figure 9.5 is as a fixed-width format file. For example, the date values always reside in the first 12 characters of each line and the temperature values are always between character 24 and character 28. This means that we could also read the file using `read.fwf()`, as shown below.

```
R> pointnemofwf <-
       read.fwf(file.path("LAS", "pointnemotemp.txt"),
                skip=8,
                widths=c(-1, 11, -11, 5),
                colClasses=c("character", "numeric"),
                col.names=c("date", "temp"))
```

```
R> pointnemofwf
```

```
         date  temp
1 16-JAN-1994 278.9
2 16-FEB-1994 280.0
3 16-MAR-1994 278.9
4 16-APR-1994 278.9
5 16-MAY-1994 277.8
6 16-JUN-1994 276.1
...
```

Again, the result is a data frame. As for the call to `read.table()`, we have specified the data type for each column, via the `colClasses` argument, and a name for each column, via `col.names`.

The `widths` argument specifies how wide each column of data is, with negative values used to ignore the specified number of characters. In this case, we have ignored the very first character on each line, we treat the next 11 characters as a `date` value, we ignore characters 13 to 23, and the final 5 characters are treated as the `temp` value.

The dates could be converted from character values to dates in exactly the same way as before.

A readLines() example

The two examples so far have demonstrated reading in the raw data for this data set, but so far we have completely ignored all of the metadata in the head of the file. This information is also very important and we would like to have some way to access it.

The readLines() function can help us here, at least in terms of getting raw text into R. The following code reads the first eight lines of the text file into a character vector.

```
R> readLines(file.path("LAS", "pointnemotemp.txt"),
             n=8)
```

```
[1] "             VARIABLE : Mean TS from clear sky composite (kelvin)"
[2] "             FILENAME : ISCCPMonthly_avg.nc"
[3] "             FILEPATH : /usr/local/fer_data/data/"
[4] "             SUBSET   : 48 points (TIME)"
[5] "             LONGITUDE: 123.8W(-123.8)"
[6] "             LATITUDE : 48.8S"
[7] "                      123.8W "
[8] "                       23"
```

Section 9.9 will describe some tools that could be used to extract the metadata *values* from this text.

A write.csv() example

As a simple demonstration of the use of functions that can *write* plain text files, we will now export the R data frame, pointnemotemp, to a new CSV file. This will create a much tidier plain text file that contains just the date and temperature values. Creating such a file is sometimes a necessary step in preparing a data set for analysis with a specific piece of analysis software.

The following code uses the write.csv() function to create a file called "pointnemoplain.csv" that contains the Point Nemo data (see Figure 9.6).

```
R> write.csv(pointnemodelim, "pointnemoplain.csv",
             quote=FALSE, row.names=FALSE)
```

The first argument in this function call is the data frame of values. The second argument is the name of the file to create.

```
date,temp
 16-JAN-1994,278.9
 16-FEB-1994,280
 16-MAR-1994,278.9
 16-APR-1994,278.9
 16-MAY-1994,277.8
 16-JUN-1994,276.1
 ...
```

Figure 9.6: The first few lines of the plain text output from the Live Access Server for the surface temperature at Point Nemo in Comma-Separated Value (CSV) format. This is a reproduction of Figure 5.4.

The `quote` argument controls whether quote-marks are printed around character values and the `row.names` argument controls whether an extra column of unique names is printed at the start of each line. In both cases, we have turned these features off by specifying the value `FALSE`.

We will continue to use the Point Nemo data set, in various formats, throughout the rest of this section.

9.7.5 Binary formats

As discussed in Section 5.3, it is only possible to extract data from a binary file format with an appropriate piece of software that understands the particular binary format.

A number of R packages exist for reading particular binary formats. For example, the **foreign** package contains functions for reading files produced by other popular statistics software systems, such as SAS, SPSS, Systat, Minitab, and Stata. As an example of support for a more general binary format, the **ncdf** package provides functions for reading netCDF files.

We will look again at the Point Nemo temperature data (see Section 1.1), this time in a netCDF format, to demonstrate a simple use of the **ncdf** package.

The following code loads the **ncdf** package and reads the file `pointnemotemp.nc`.

```
R> library("ncdf")
R> nemonc <- open.ncdf(file.path("LAS",
                                  "pointnemotemp.nc"))
```

One difference with this example compared to the functions in the previous section is that the result of reading the netCDF file is *not* a data frame.

```
R> class(nemonc)

[1] "ncdf"
```

This data structure is essentially a list that contains the information from the netCDF file. We can extract components of this list by hand, but a more convenient approach is to use several other functions provided by the **ncdf** package to extract the most important pieces of information from the `ncdf` data structure.

If we display the `nemonc` data structure, we can see that the file contains a single variable called `Temperature`.

```
R> nemonc

file LAS/pointnemotemp.nc has 1 dimensions:
Time   Size: 48
------------------------
file LAS/pointnemotemp.nc has 1 variables:
double Temperature[Time]  Longname:Temperature
```

We can extract that variable from the file with the function `get.var.ncdf()`.

```
R> nemoTemps <- get.var.ncdf(nemonc, "Temperature")
R> nemoTemps

 [1] 278.9 280.0 278.9 278.9 277.8 276.1 276.1 275.6 275.6
[10] 277.3 276.7 278.9 281.6 281.1 280.0 278.9 277.8 276.7
[19] 277.3 276.1 276.1 276.7 278.4 277.8 281.1 283.2 281.1
[28] 279.5 278.4 276.7 276.1 275.6 275.6 276.1 277.3 278.9
[37] 280.5 281.6 280.0 278.9 278.4 276.7 275.6 275.6 277.3
[46] 276.7 278.4 279.5
```

The netCDF file of Point Nemo data also contains information about the date that each temperature value corresponds to. This variable is called `"Time"`. The following code reads this information from the file.

```
R> nemoTimes <- get.var.ncdf(nemonc, "Time")
R> nemoTimes

 [1]  8781  8812  8840  8871  8901  8932  8962  8993  9024
[10]  9054  9085  9115  9146  9177  9205  9236  9266  9297
[19]  9327  9358  9389  9419  9450  9480  9511  9542  9571
[28]  9602  9632  9663  9693  9724  9755  9785  9816  9846
[37]  9877  9908  9936  9967  9997 10028 10058 10089 10120
[46] 10150 10181 10211
```

Unfortunately, these do not look very much like dates.

This demonstrates that, even with binary file formats, it is often necessary to coerce data from the storage format that has been used in the file to a more convenient format for working with the data in RAM.

The netCDF format only allows numbers or text to be stored, so this date information has been stored in the file as numbers. However, additional information, metadata, can be stored along with the variable data in a netCDF file; netCDF calls this additional information "attributes". In this case, the meaning of these numbers representing dates has been stored as the "units" attribute of this variable and the following code uses the `att.get.ncdf()` function to extract that attribute from the file.

```
R> att.get.ncdf(nemonc, "Time", "units")

$hasatt
[1] TRUE

$value
[1] "number of days since 1970-01-01"
```

This tells us that the dates were stored as a number of days since January 1st 1970. Using this information, we can convert the numbers from the file back into real dates with the `as.Date()` function. The `origin` argument allows us to specify the meaning of the numbers.

```
R> nemoDates <- as.Date(nemoTimes, origin="1970-01-01")
```

```
R> nemoDates
```

```
 [1] "1994-01-16" "1994-02-16" "1994-03-16" "1994-04-16"
 [5] "1994-05-16" "1994-06-16" "1994-07-16" "1994-08-16"
 [9] "1994-09-16" "1994-10-16" "1994-11-16" "1994-12-16"
...
```

In most cases, where a function exists to read a particular binary format, there will also be a function to write data out in that format. For example, the **ncdf** package also provides functions to save data from RAM to an external file in netCDF format.

9.7.6 Spreadsheets

When data is stored in a spreadsheet, one common approach is to save the data in a text format as an intermediate step and then read the text file into R using the functions from Section 9.7.3.

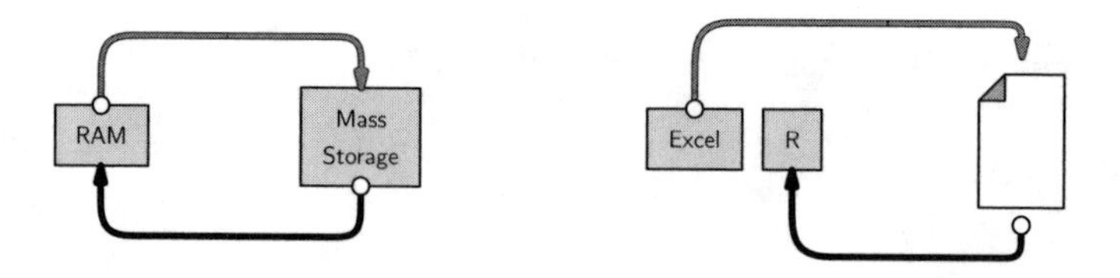

This makes the data easy to share, because text formats are very portable, but it has the disadvantage that another copy of the data is created.

This is less efficient in terms of storage space and it creates issues if the original spreadsheet is updated.

If changes are made to the original spreadsheet, at best, there is extra work to be done to update the text file as well. At worst, the text file is forgotten and the update does not get propagated to other places.

pointnemotemp.xls

	A	B	C	D	E	F	G	H
1	16-Jan-94	278.9						
2	16-Feb-94	280.0						
3	16-Mar-94	278.9						
4	16-Apr-94	278.9						
5	16-May-94	277.8						
6	16-Jun-94	276.1						
7	16-Jul-94	276.1						
8	16-Aug-94	275.6						
9	16-Sep-94	275.6						
10	16-Oct-94	277.3						

temperatures

Figure 9.7: Part of the Excel spreadsheet containing the surface temperatures at Point Nemo.

There are several packages that provide ways to directly read data from a spreadsheet into R. One example is the (Windows only) **xlsReadWrite** package, which includes the `read.xls()` function for reading data from an Excel spreadsheet.

Figure 9.7 shows a screen shot of the Point Nemo temperature data (see Section 1.1) stored in a Microsoft Excel spreadsheet.

These data can be read into R using the following code.

```
R> library("xlsReadWrite")
R> read.xls("temperatures.xls", colNames=FALSE)
```

```
     V1    V2
1 34350 278.9
2 34381 280.0
3 34409 278.9
4 34440 278.9
5 34470 277.8
6 34501 276.1
...
```

Notice that the date information has come across as numbers. This is another example of the type coercion that can easily occur when transferring between different formats.

As before, we can easily convert the numbers to dates if we know the reference date for these numbers. Excel represents dates as the number of days since the 0th of January 1900, so we can recover the real dates with the following code.

```
R> dates <- as.Date(temps$V1 - 2, origin="1900-01-01")
```

```
R> dates
```

```
 [1] "1994-01-16" "1994-02-16" "1994-03-16" "1994-04-16"
 [5] "1994-05-16" "1994-06-16" "1994-07-16" "1994-08-16"
 [9] "1994-09-16" "1994-10-16" "1994-11-16" "1994-12-16"
...
```

We have to subtract 2 in this calculation because the Excel count starts from the 0th rather than the 1st of January *and* because Excel thinks that 1900 was a leap year (apparently to be compatible with the Lotus 123 spreadsheet software). Sometimes, computer technology is not straightforward.

The **gdata** package provides another way to access Excel spreadsheets with its own `read.xls()` function and it is also possible to access Excel spreadsheets via a technology called ODBC (see Section 9.7.8).

9.7.7 XML

In this section, we look at how to get information that has been stored in an XML document into R.

Although XML files are plain text files, functions like `read.table()` from Section 9.7.3 are of no use because they only work with data that are arranged in a plain text *format*, with the data laid out in rows and columns within the file.

It is possible to read an XML document into R as a character vector using a standard function like `readLines()`. However, extracting the information from the text is not trivial because it requires knowledge of XML.

Fortunately, there is an R package called **XML** that contains functions for

```
<?xml version="1.0"?>
<temperatures>
    <variable>Mean TS from clear sky composite (kelvin)</variable>
    <filename>ISCCPMonthly_avg.nc</filename>
    <filepath>/usr/local/fer_dsets/data/</filepath>
    <subset>93 points (TIME)</subset>
    <longitude>123.8W(-123.8)</longitude>
    <latitude>48.8S</latitude>
    <case date="16-JAN-1994" temperature="278.9" />
    <case date="16-FEB-1994" temperature="280" />
    <case date="16-MAR-1994" temperature="278.9" />
    <case date="16-APR-1994" temperature="278.9" />
    <case date="16-MAY-1994" temperature="277.8" />
    <case date="16-JUN-1994" temperature="276.1" />

    ...

</temperatures>
```

Figure 9.8: The first few lines of the surface temperature at Point Nemo in an XML format. This is a reproduction of Figure 5.16.

reading and extracting data from XML files into R.

We will use the Point Nemo temperature data, in an XML format, to demonstrate some of the functions from the **XML** package. Figure 9.8 shows one possible XML format for the the Point Nemo temperature data.

There are several approaches to working with XML documents in R using the **XML** package, but in this section, we will only consider the approach that allows us to use XPath queries (see Section 7.3.1).

The first thing to do is to read the XML document into R using the function `xmlTreeParse()`.

```
R> library("XML")
```

```
R> nemoInternalDoc <-
       xmlTreeParse(file.path("LAS",
                              "pointnemotemp.xml"),
                    useInternalNodes=TRUE)
```

The first argument to this function is the name and location of the XML file. The useInternalNodes argument is necessary when we want to use XPath to select elements from the file.

It is important to point out that the data structure that is created in RAM by this code, nemoInternalDoc, is *not* a data frame.

```
R> class(nemoInternalDoc)
```

```
[1] "XMLInternalDocument"
```

We must use other special functions from the **XML** package to work with this data structure.

In particular, the xpathSApply() function allows us to select elements from this data structure using XPath expressions.

In the following example, we extract the temperature attribute values from all case elements in the XML document. The XPath expression "/temperatures/case/@temperature" selects all of the temperature attributes of the case elements within the root temperatures element.

```
R> nemoDocTempText <-
       xpathSApply(nemoInternalDoc,
                   "/temperatures/case/@temperature",
                   xmlValue)
```

```
R> nemoDocTempText
```

```
 [1] "278.9" "280"   "278.9" "278.9" "277.8" "276.1" "276.1"
 [8] "275.6" "275.6" "277.3" "276.7" "278.9" "281.6" "281.1"
[15] "280"   "278.9" "277.8" "276.7" "277.3" "276.1" "276.1"
...
```

The first argument to xpathSApply() is the data structure previously created by xmlTreeParse(), the second argument is the XPath expression, and the third argument is a function that is called for each temperature attribute. The xmlValue() function is used here to get the value of the temperature attribute. This function call may be easier to understand once we have seen the "apply" functions in Section 9.8.7.

One important point about the above result, nemoDocTempText, is that it is a character vector. This reflects the fact that everything is stored as text within an XML document. If we want to have numeric values to work with, we need to coerce these text values into numbers.

Before we do that, the following code shows how to extract the date values from the XML document as well. The only difference from the previous call is the XPath that we use.

```
R> nemoDocDateText <-
       xpathSApply(nemoInternalDoc,
                   "/temperatures/case/@date",
                   xmlValue)
```

```
R> nemoDocDateText
```

```
 [1] "16-JAN-1994" "16-FEB-1994" "16-MAR-1994" "16-APR-1994"
 [5] "16-MAY-1994" "16-JUN-1994" "16-JUL-1994" "16-AUG-1994"
 [9] "16-SEP-1994" "16-OCT-1994" "16-NOV-1994" "16-DEC-1994"
...
```

Again, the values are all text, so we need to coerce them to dates. The following code performs the appropriate type coercions and combines the dates and temperatures into a data frame.

```
R> data.frame(date=as.Date(nemoDocDateText, "%d-%b-%Y"),
              temp=as.numeric(nemoDocTempText))
```

```
        date  temp
1 1994-01-16 278.9
2 1994-02-16 280.0
3 1994-03-16 278.9
4 1994-04-16 278.9
5 1994-05-16 277.8
6 1994-06-16 276.1
...
```

With this approach to reading XML files, there is one final step: we need to signal that we are finished with the file by calling the free() function.

```
R> free(nemoInternalDoc)
```

9.7.8 Databases

Very large data sets are often stored in relational databases. As with spreadsheets, a simple approach to extracting information from the database is to export it from the database to text files and work with the text files. This is an even worse option for databases than it was for spreadsheets because it is more common to extract just part of a database, rather than an entire spreadsheet. This can lead to several different text files from a single database, and these are even harder to maintain if the database changes.

A superior option is to extract information directly from the database management system into R.

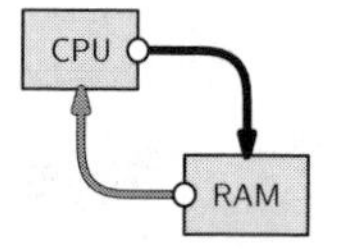

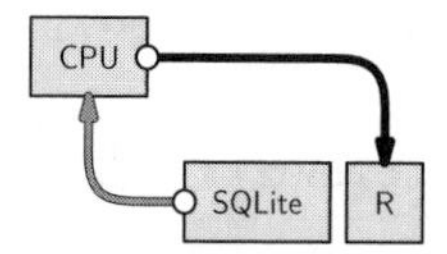

There are packages for connecting directly to several major database management systems. Two main approaches exist, one based on the **DBI** package and one based on the **RODBC** package.

The **DBI** package defines a set of standard (generic) functions for communicating with a database, and a number of other packages, e.g., **RMySQL** and **RSQLite**, build on that to provide functions specific to a particular database system. The important functions to know about with this approach are:

`dbDriver()`
: to create a "device driver", which contains information about a particular database management system.

`dbConnect(drv)`
: to create a "connection" to a database. Requires a device driver, `drv`, as created by `dbDriver()`.

`dbGetQuery(conn, statement)`
: to send the SQL command, `statement`, to the database and receive a result. The result is a data frame. Requires a connection, `conn`, as created by `dbConnect()`.

`dbDisconnect(conn)`
: to sever the connection with the database and release resources.

The **RODBC** package defines functions for communicating with any ODBC

(Open Database Connectivity) compliant software. This allows connections with many different types of software, including, but not limited to, most database management systems. For example, this approach can also be used to extract information from a Microsoft Excel spreadsheet.

The important functions to know about with this approach are:

`odbcConnect()`
: to connect to the ODBC application.

`sqlQuery(channel, query)`
: to send an SQL command to the database and receive a result, as a data frame. Requires a connection, `channel`, that has been created by `odbcConnect()`.

`odbcClose(channel)`
: to sever the ODBC connection and release resources.

The **RODBC** approach makes it possible to connect to a wider range of other software systems, but it may involve installation of additional software.

The simplest approach of all is provided by the **RSQLite** package because it includes the complete SQLite application, so no other software needs to be installed. However, this will only be helpful if the data are stored in an SQLite database.

The next section demonstrates an example usage of the **RSQLite** package.

9.7.9 Case study: The Data Expo (continued)

The Data Expo data set (see Section 5.2.8) contains several different atmospheric measurements, all measured at 72 different time periods and 576 different locations. These data have been stored in an SQLite database, with a table for location information, a table for time period information, and a table of the atmospheric measurements (see Section 7.1).

The following SQL code extracts information for the first two locations in the location table.

```
SELECT *
    FROM location_table
    WHERE ID = 1 OR ID = 2;
```

The following code carries out this query from within R. The first step is to connect to the database.

```
R> library("RSQLite")
```

```
R> con <- dbConnect(dbDriver("SQLite"),
                    dbname="NASA/dataexpo")
```

Having established the connection, we can send SQL code to the DBMS.

```
R> result <-
       dbGetQuery(con,
                  "SELECT *
                       FROM location_table
                       WHERE ID = 1 OR ID = 2")
R> result
```

```
  ID longitude latitude elevation
1  1   -113.75    36.25   1526.25
2  2   -111.25    36.25   1759.56
```

Notice that the result is a *data frame*. The final step is to release our connection to SQLite.

```
R> dbDisconnect(con)
```

```
[1] TRUE
```

Recap

There are functions for reading plain text files in both delimited and fixed-width formats. These functions create a data frame from the data values in the file.

For many binary formats and spreadsheets, there exist packages with special functions for reading files in the relevant format.

The **XML** *package provides special functions for working with XML documents.*

Several packages provide special functions for extracting data from relational databases. The result of a query to a database is a data frame.

9.8 Data manipulation

This section describes a number of techniques for rearranging objects in R, particularly larger data structures, such as data frames, matrices, and lists.

Some of these techniques are useful for basic exploration of a data set. For example, we will look at functions for sorting data and for generating tables of counts.

Other techniques will allow us to perform more complex tasks such as restructuring data sets to convert them from one format to another, splitting data structures into smaller pieces, and combining smaller pieces to form larger data structures.

This section is all about starting with a data structure in RAM and calculating new values, or just rearranging the existing values, to generate a new data structure in RAM.

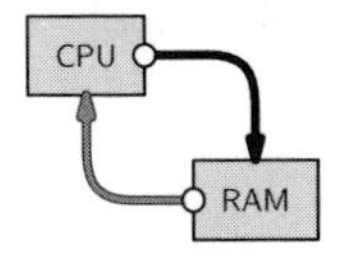

In order to provide a motivation for some of these techniques, we will use the following case study throughout this section.

9.8.1 Case study: New Zealand schools

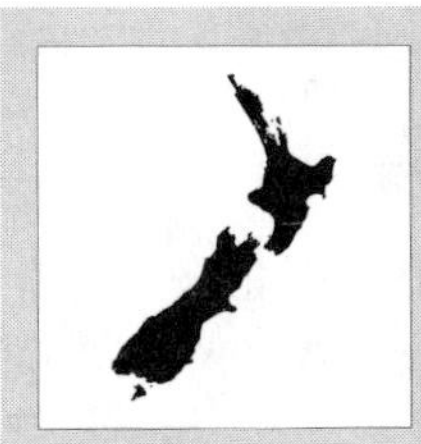

New Zealand consists of the North Island, the South Island, and Stewart Island (all visible in the image to the left), plus over 100 other islands scattered over the Pacific Ocean and the Southern Ocean.

The New Zealand Ministry of Education provides basic information for all primary and secondary schools in the country.[2] In this case study, we will work with a subset of this information that contains the following variables:

[2]The data were originally obtained from `http://www.minedu.govt.nz/web/downloadable/dl6434_v1/directory-schools-web.xls` and similar files are now available from `http://www.educationcounts.govt.nz/statistics/tertiary_education/27436`.

```
"ID","Name","City","Auth","Dec","Roll"
1015,"Hora Hora School","Whangarei","State",2,318
1052,"Morningside School","Whangarei","State",3,200
1062,"Onerahi School","Whangarei","State",4,455
1092,"Raurimu Avenue School","Whangarei","State",2,86
1130,"Whangarei School","Whangarei","State",4,577
1018,"Hurupaki School","Whangarei","State",8,329
1029,"Kamo Intermediate","Whangarei","State",5,637
1030,"Kamo School","Whangarei","State",5,395
...
```

Figure 9.9: The first few lines of the file `schools.csv`. This file contains information for all primary and secondary schools in New Zealand, in a CSV format.

ID
A unique numeric identifier for each school.

Name
The name of the school.

City
The city where the school is located.

Auth
The "authority" or ownership of the school. Possible values are `Private`, `State`, `State Integrated`, and `Other`. A state integrated school is one that was private in the past, but is now state-owned; these schools usually have some special character (e.g., an emphasis on a particular religious belief).

Dec
The "decile" of the school, which is a measure of the socio-economic status of the families of the students of the school. A lower decile roughly corresponds to a poorer school population. The value ranges from 1 (poorest) to 10 (richest).

Roll
The number of students enrolled at the school (as of July 2007).

These data have been stored in CSV format in a file called `schools.csv` (see Figure 9.9).

Using functions from the last section, we can easily read this information into R and store it as a data frame called `schools`. The `as.is` argument is used to ensure that the school names are kept as text and are not treated as a factor.

```
R> schools <-
       read.csv(file.path("Schools", "schools.csv"),
                as.is=2)
```

```
R> schools
```

```
    ID                    Name      City  Auth Dec Roll
1 1015       Hora Hora School Whangarei State   2  318
2 1052     Morningside School Whangarei State   3  200
3 1062         Onerahi School Whangarei State   4  455
4 1092 Raurimu Avenue School Whangarei State   2   86
5 1130       Whangarei School Whangarei State   4  577
6 1018        Hurupaki School Whangarei State   8  329
...
```

In cases like this, where the data set is much too large to view all at once, we can make use of functions to explore the basic characteristics of the data set. For example, it is useful to know just how large the data set is. The following code tells us the number of rows and columns in the data frame.

```
R> dim(schools)
```

```
[1] 2571    6
```

There are 6 variables measured on 2,571 schools.

We will now go on to look at some data manipulation tasks that we could perform on this data set.

9.8.2 Transformations

A common task in data preparation is to create new variables from an existing data set. As an example, we will generate a new variable that distinguishes between "large" schools and "small" schools.

This new variable will be based on the median school roll. Any school with a roll greater than the median will be "large".

```
R> rollMed <- median(schools$Roll)
R> rollMed
```

```
[1] 193
```

We can use the `ifelse()` function to generate a character vector recording `"large"` and `"small"`. The first argument to this function, called `test`, is a logical vector. The second argument, called `yes`, provides the result wherever the `test` argument is `TRUE` and the third argument, called `no`, provides the result wherever the `test` argument is `FALSE`.

```
R> size <- ifelse(test=schools$Roll > rollMed,
                  yes="large", no="small")
```

```
R> size
```

```
 [1] "large" "large" "large" "small" "large" "large" "large"
 [8] "large" "large" "large" "large" "large" "small" "large"
...
```

The diagram below illustrates that the `ifelse()` function selects elements from the `yes` argument when the `test` argument is `TRUE` and from the `no` argument when `test` is `FALSE`.

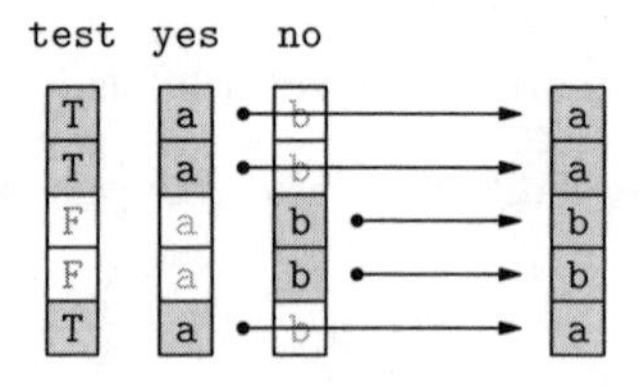

We will now add this new variable to the original data frame to maintain the correspondence between the school size labels and the original rolls that they were derived from.

This can be done in several ways. The simplest is to assign a new variable to the data frame, as shown in the following code. The data frame does not have an existing column called `Size`, so this assignment adds a new column.

```
R> schools$Size <- size
```

```
R> schools
```

```
    ID                  Name      City  Auth Dec Roll  Size
1 1015      Hora Hora School Whangarei State   2  318 large
2 1052    Morningside School Whangarei State   3  200 large
3 1062        Onerahi School Whangarei State   4  455 large
4 1092 Raurimu Avenue School Whangarei State   2   86 small
5 1130      Whangarei School Whangarei State   4  577 large
6 1018       Hurupaki School Whangarei State   8  329 large
...
```

Another approach, which gives the same result, is to use the `transform()` function.

```
R> schools <- transform(schools, Size=size)
```

If we want to remove a variable from a data frame, we can assign the value `NULL` to the appropriate column.

```
R> schools$size <- NULL
```

Alternatively, we can use subsetting to retain certain columns or leave out certain columns. This approach is better if more than one column needs to be removed. For example, the following code removes the new, seventh column from the data frame.

```
R> schools <- schools[, -7]
```

```
R> schools
```

```
    ID                     Name      City  Auth Dec Roll
1 1015        Hora Hora School Whangarei State   2  318
2 1052      Morningside School Whangarei State   3  200
3 1062          Onerahi School Whangarei State   4  455
4 1092 Raurimu Avenue School Whangarei State   2   86
5 1130        Whangarei School Whangarei State   4  577
6 1018         Hurupaki School Whangarei State   8  329
...
```

Binning

The previous example converted a numeric variable into two categories, small and large. The more general case is to convert a numeric variable into any number of categories.

For example, rather than just dividing the schools into large and small, we could group the schools into five different size categories. This sort of transformation is possible using the `cut()` function.

```
R> rollSize <-
       cut(schools$Roll, 5,
           labels=c("tiny", "small", "medium",
                    "large", "huge"))
```

The first argument to this function is the numeric vector to convert. The second argument says that the range of the numeric vector should be broken into five equal-sized intervals. The schools are then categorized according to which interval their roll lies within. The `labels` argument is used to provide levels for the resulting factor.

Only the first few schools are shown below and they are all tiny; the important point is that the result is a factor with five levels.

```
R> head(rollSize)
```

```
[1] tiny tiny tiny tiny tiny tiny
Levels: tiny small medium large huge
```

A better view of the result can be obtained by counting how many schools there are of each size. This is what the following code does; we will see more about functions to create tables of counts in Section 9.8.4.

```
R> table(rollSize)

rollSize
  tiny  small medium  large   huge
  2487     75      8      0      1
```

9.8.3 Sorting

Another common task that we can perform is to sort a set of values into ascending or descending order.

In R, the function `sort()` can be used to arrange a vector of values in order, but of more general use is the `order()` function, which returns the indices of the sorted values.

As an example, the information on New Zealand schools is roughly ordered by region, from North to South in the original file. We might want to order the schools by size instead.

The following code sorts the `Roll` variable by itself.

```
R> sort(schools$Roll)

 [1] 5 5 6 6 6 7 7 7 7 8 9 9 9 9 9 9 9 ...
```

There are clearly some very small schools in New Zealand.

It is also easy to sort in decreasing order, which reveals that the largest school is the largest by quite some margin.

```
R> sort(schools$Roll, decreasing=TRUE)

 [1] 5546 3022 2613 2588 2476 2452 2417 2315 2266 2170 ...
```

However, what would be much more useful would be to know which schools these are. That is, we would like to sort not just the school rolls, but the entire schools data frame.

To do this, we use the `order()` function, as shown in the following code.

```
R> rollOrder <- order(schools$Roll, decreasing=TRUE)
```

```
R> rollOrder
```

```
 [1] 1726  301  376 2307  615  199  467  373  389  241 ...
```

This result says that, in order to sort the data frame in descending roll order, we should use row 1726 first, then row 301, then row 376, and so on.

These values can be used as indices for subsetting the entire `schools` data frame, as shown in the following code. Recall that, by only specifying a row index and by leaving the column index blank, we will get all columns of the `schools` data frame.

```
R> schools[rollOrder, ]
```

```
      ID                    Name         City  Auth Dec Roll
1726 498    Correspondence School   Wellington State  NA 5546
301   28        Rangitoto College     Auckland State  10 3022
376   78         Avondale College     Auckland State   4 2613
2307 319      Burnside High School Christchurch State   8 2588
615   41         Macleans College     Auckland State  10 2476
199   43        Massey High School     Auckland State   5 2452
467   54         Auckland Grammar     Auckland State  10 2417
373   69 Mt Albert Grammar School     Auckland State   7 2315
389   74      Mt Roskill Grammar     Auckland State   4 2266
...
```

The largest body of New Zealand school students represents those gaining public education via correspondence (from the governmental base in Wellington), but most of the other large schools are in Auckland, which is the largest city in New Zealand.

The other advantage of using the `order()` function is that more than one vector of values may be given and any ties in the first vector will be broken by ordering on the second vector. For example, the following code sorts the rows of the `schools` data frame first by city and *then* by number of students (in decreasing order). In the case of the `City` variable, which is a character vector, the order is alphabetic. Again, we are specifying `decreasing=TRUE` to get descending order.

```
R> schools[order(schools$City, schools$Roll,
                 decreasing=TRUE), ]
```

```
       ID                      Name       City  Auth Dec Roll
2548  401           Menzies College    Wyndham State   4  356
2549 4054            Wyndham School    Wyndham State   5   94
1611 2742          Woodville School  Woodville State   3  147
1630 2640           Papatawa School  Woodville State   7   27
2041 3600            Woodend School    Woodend State   9  375
1601  399 Central Southland College     Winton State   7  549
...
```

The first two schools are both in Wyndham, with the larger school first and the smaller school second, then there are two schools from Woodville, larger first and smaller second, and so on.

9.8.4 Tables of counts

Continuing our exploration of the New Zealand schools data set, we might be interested in how many schools are private and how many are state-owned. This is an example where we want to obtain a **table of counts** for a categorical variable. The function `table()` may be used for this task.

```
R> authTable <- table(schools$Auth)
```

```
R> authTable
```

```
           Other          Private            State State Integrated
               1               99             2144              327
```

This result shows the number of times that each different value occurs in the `Auth` variable. As expected, most schools are public schools, owned by the state.

As usual, we should take notice of the data structure that this function has returned.

```
R> class(authTable)
```

```
[1] "table"
```

This is not one of the basic data structures that we have focused on. However, tables in R behave very much like arrays, so we can use what we already know about working with arrays. For example, we can subset a table just like we subset an array. If we need to, we can also convert a table to an array, or even to a data frame.

As a brief side track, the table of counts above also shows that there is only one school in the category "Other". How do we find that school? One approach is to generate a logical vector and use that to subscript the data frame, as in the following code.

```
R> schools[schools$Auth == "Other", ]
```

```
      ID             Name         City  Auth Dec Roll
2315 518 Kingslea School Christchurch Other   1   51
```

It turns out that this school is not state-owned, but still receives its funding from the government because it provides education for students with learning or social difficulties.

Getting back to tables of counts, another question we might ask is how the school decile relates to the school ownership. If we give the `table()` function more than one argument, it will cross-tabulate the arguments, as the following code demonstrates.

```
R> table(Dec=schools$Dec, Auth=schools$Auth)
```

```
    Auth
Dec  Other Private State State Integrated
  1      1       0   259               12
  2      0       0   230               22
  3      0       2   208               35
  4      0       6   219               28
  5      0       2   214               38
  6      0       2   215               34
  7      0       6   188               45
  8      0      11   200               45
  9      0      12   205               37
  10     0      38   205               31
```

Again, the result of this function call is a table data structure. This time it is a two-dimensional table, which is very similar to a two-dimensional array.

In this example, we have provided names for the arguments, Dec and Auth, and these have been used to provide overall labels for each dimension of the table.

This table has one row for each decile and one column for each type of ownership. For example, the third row of the table tells us that there are 2 private schools, 209 state schools, and 35 state integrated schools with a decile of 3.

The most obvious feature of this result is that private schools tend to be in wealthier areas, with higher deciles.

Another function that can be used to generate tables of counts is the xtabs() function. This is very similar to the table() function, except that the factors to cross-tabulate are specified in a formula, rather than as separate arguments.

The two-dimensional table above could also be generated with the following code.

```
R> xtabs( ~ Dec + Auth, data=schools)
```

One advantage of this approach is that the symbols used in the formula are automatically found in the data frame provided in the data argument, so, for example, there is no need to specify the Auth variable as schools$Auth, as we had to do in the previous call to the table() function.

9.8.5 Aggregation

R provides many functions for calculating numeric summaries of data. For example, the min() and max() functions calculate minimum and maximum values for a vector, sum() calculates the sum, and the mean() function calculates the average value. The following code calculates the average value of the number of students enrolled at New Zealand schools.

```
R> mean(schools$Roll)

[1] 295.4737
```

This "grand mean" value is not very interesting by itself. What would be more interesting would be to calculate the average enrolment for different types of schools and compare, for example, state schools and private schools.

We could calculate these values individually, by extracting the relevant subset of the data. For example, the following code extracts just the enrolments for private schools and calculates the mean for those values.

```
R> mean(schools$Roll[schools$Auth == "Private"])

[1] 308.798
```

However, a better way to work is to let R determine the subsets and calculate all of the means in one go. The `aggregate()` function provides one way to perform this sort of task.

The following code uses `aggregate()` to calculate the average enrolment for New Zealand schools broken down by school ownership.

There are four different values of `schools$Auth`, so the result is four averages. We can easily check the answer for `Other` schools because there is only one such school and we saw on page 296 that this school has a roll of 51.

```
R> aggregate(schools["Roll"],
             by=list(Ownership=schools$Auth),
             FUN=mean)

         Ownership     Roll
1            Other  51.0000
2          Private 308.7980
3            State 300.6301
4 State Integrated 258.3792
```

This result shows that the average school size is remarkably similar for all types of school ownership (ignoring the `"Other"` case because there is only one such school).

The `aggregate()` function can take a bit of getting used to, so we will take a closer look at the arguments to this function.

There are three arguments to the `aggregate()` call. The first argument provides the values that we want to average. In this case, these values are the enrolments for each school. A minor detail is that we have provided a *data frame* with only one column, the `Roll` variable, rather than just a vector. It is possible to provide either a vector or a data frame, but the advantage of using a data frame is that the second column in the result has a sensible name.

We could have used `schools$Roll`, a vector, instead of `schools["Roll"]`, a data frame, but then the right-hand column of the result would have had an uninformative name, `x`.

The second argument to the `aggregate()` function is called `by` and the value must be a *list*. In this case, there is only one component in the list, the `Auth` variable, which describes the ownership of each school. This argument is used to generate subsets of the first argument. In this case, we subset the school enrolments into four groups, corresponding to the four different types of school ownership. By providing a name for the list component, `Ownership`, we control the name of the first column in the result. If we had not done this, then the first column of the result would have had a much less informative name, `Group.1`.

The third argument to the `aggregate()` function, called `FUN`, is the name of a function. This function is called on each subset of the data. In this case, we have specified the `mean()` function, so we have asked for the average enrolments to be calculated for each different type of school ownership. It is possible to specify any function name here, as long as the function returns only a single value as its result. For example, `sum()`, `min()`, and `max()` are all possible alternatives.

The diagram below provides a conceptual view of how `aggregate()` works.

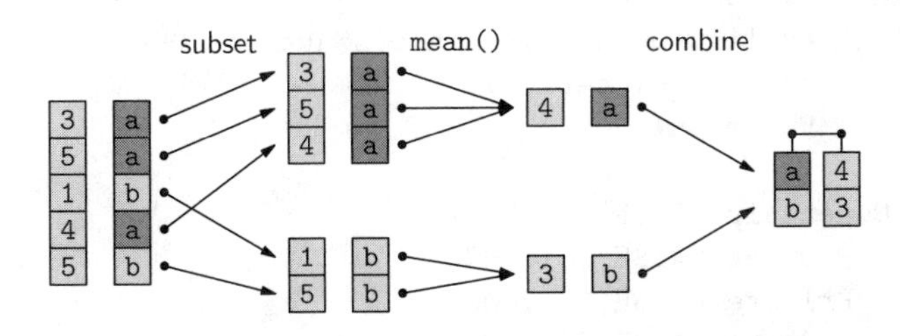

In the terminology of data queries in Section 7.2.1, the `aggregate()` function acts very much like an SQL query with a `GROUP BY` clause.

Another point to make about the `aggregate()` function is that the value returned by the function is a *data frame*. This is convenient because data frames are easy to understand and we have lots of tools for working with data frames.

We will now look at a slightly more complicated use of the aggregate() function. For this example, we will generate a new vector called rich that records whether each school's decile is greater than 5. The following code creates the new vector.

```
R> rich <- schools$Dec > 5
```

```
R> rich
```

```
[1] FALSE FALSE FALSE FALSE FALSE  TRUE FALSE ...
```

The vector rich provides a crude measure of whether each school is in a wealthy area or not. We will now use this variable in a more complicated call to aggregate().

Because the by argument to aggregate() is a list, it is possible to provide more than one factor. This means that we can produce a result for all possible combinations of two or more factors. In the following code, we provide a list of *two* factors as the by argument, so the result will be the average enrolments broken down by *both* the type of school ownership *and* whether the school is in a wealthy area.

```
R> aggregate(schools["Roll"],
             by=list(Ownership=schools$Auth,
                     Rich=rich),
             FUN=mean)
```

```
         Ownership  Rich     Roll
1            Other FALSE  51.0000
2          Private FALSE 151.4000
3            State FALSE 261.7487
4 State Integrated FALSE 183.2370
5          Private  TRUE 402.5362
6            State  TRUE 338.8243
7 State Integrated  TRUE 311.2135
```

The result of the aggregation is again a data frame, but this time there are three columns. The first two columns indicate the different groups and the third gives the average for each group.

The result suggests that, on average, schools in wealthier areas have more students.

One limitation of the `aggregate()` function is that it only works with functions that return a single value. If we want to calculate, for example, the *range* of enrolments for each type of school—the minimum and the maximum together—the `range()` function will perform the calculation for us, but we cannot use `range()` with `aggregate()` because it returns two values. Instead, we need to use the `by()` function.

The following code calculates the range of enrolments for *all* New Zealand schools, demonstrating that the result is a numeric vector containing *two* values, 5 and 5546.

```
R> range(schools$Roll)

[1]    5 5546
```

The following code uses the `by()` function to generate the range of enrolments broken down by school ownership.

```
R> rollRanges <-
       by(schools["Roll"],
          INDICES=list(Ownership=schools$Auth),
          FUN=range)
R> rollRanges

Ownership: Other
[1] 51 51
---------------------------------------------
Ownership: Private
[1]    7 1663
---------------------------------------------
Ownership: State
[1]    5 5546
---------------------------------------------
Ownership: State Integrated
[1]   18 1475
```

The arguments to the `by()` function are very similar to the arguments for `aggregate()` (with some variation in the argument names) and the effect is also very similar: the first argument is broken into subsets, with one subset for each different value in the second argument, and the third argument provides a function to call on each subset.

However, the result of `by()` is *not* a data frame like for `aggregate()`. It is a very different sort of data structure.

```
R> class(rollRanges)
```

```
[1] "by"
```

We have not seen this sort of data structure before, but a `by` object behaves very much like a list, so we can work with this result just like a list.

The result of the call to `by()` gives the range of enrolments for each type of school ownership. It suggests that most of the large schools are owned by the state.

9.8.6 Case study: NCEA

The concept of measuring student achievement by assigning a grade was first introduced by William Farish, a tutor at the University of Cambridge, in 1792.

In order to motivate some of the remaining sections, here we introduce another, related New Zealand schools data set for us to work with.

The National Certificates of Educational Achievement (NCEA) are used to measure students' learning in New Zealand secondary schools. Students usually attempt to achieve NCEA Level 1 in their third year of secondary schooling, Level 2 in their fourth year, and Level 3 in their fifth and final year of secondary school.

Each year, information on the percentage of students who achieved each NCEA level is reported for all New Zealand secondary schools. In this case study, we will look at NCEA achievement percentages for 2007.

The data are stored in a plain text, colon-delimited format, in a file called `NCEA2007.txt`. There are four columns of data: the school name, plus the

```
Name:Level1:Level2:Level3
Al-Madinah School:61.5:75:0
Alfriston College:53.9:44.1:0
Ambury Park Centre for Riding Therapy:33.3:20:0
Aorere College:39.5:50.2:30.6
Auckland Girls' Grammar School:71.2:78.9:55.5
Auckland Grammar:22.1:30.8:26.3
Auckland Seventh-Day Adventist H S:50.8:34.8:48.9
Avondale College:57.3:49.8:44.6
Baradene College:89.3:89.7:88.6
...
```

Figure 9.10: The first few lines of the file NCEA2007.txt. This file contains information about the percentage of students gaining NCEA qualifications at New Zealand secondary schools in 2007. This is a plain text, colon-delimited format.

three achievement percentages for the three NCEA levels. Figure 9.10 shows the first few lines of the file.

The following code uses the `read.table()` function to read these data into a data frame called `NCEA`. We have to specify the colon delimiter via `sep=":"`. Also, because some school names have an apostrophe, we need to specify `quote=""`. Otherwise, the apostrophes are interpreted as the start of a text field. The last two arguments specify that the first line of the file contains the variable names (`header=TRUE`) and that the first column of school names should be treated as character data (not as a factor).

```
R> NCEA <- read.table(file.path("Schools", "NCEA2007.txt"),
                      sep=":", quote="",
                      header=TRUE, as.is=TRUE)
```

```
R> NCEA
```

```
                                 Name Level1 Level2 Level3
1                   Al-Madinah School   61.5   75.0    0.0
2                   Alfriston College   53.9   44.1    0.0
3 Ambury Park Centre for Riding Therapy 33.3   20.0    0.0
4                      Aorere College   39.5   50.2   30.6
5       Auckland Girls' Grammar School  71.2   78.9   55.5
6                    Auckland Grammar   22.1   30.8   26.3
7   Auckland Seventh-Day Adventist H S   50.8   34.8   48.9
...
```

As with the schools data frame, the data set is too large to view all at once, so we will only show a few rows for most examples in the following sections.

The dim() function tells us that, in total, there are 88 schools and 4 variables in this data set. This data set only has Auckland schools.

```
R> dim(NCEA)

[1] 88  4
```

We will now explore this data set in order to demonstrate some more data manipulation techniques.

9.8.7 The "apply" functions

The NCEA data frame contains three columns of numeric data—the percentage of students achieving NCEA at the three different NCEA levels. Something we could quickly look at is whether the achievement percentages are similar, on average, for each NCEA level.

One way to look at these averages is to extract each column and calculate an average for each column. For example, the following code calculates the average achievement percentage for NCEA Level 1.

```
R> mean(NCEA$Level1)

[1] 62.26705
```

However, as in the aggregation case, there is a smarter way to work, which is to let R extract the columns and calculate a mean for all columns in a single function call. This sort of task is performed by the apply() function.

The following code uses the apply() function to calculate the average achievement percentage for each NCEA level.

```
R> apply(NCEA[2:4], MARGIN=2, FUN=mean)

  Level1   Level2   Level3
62.26705 61.06818 47.97614
```

The result suggests that a slightly lower percentage of students achieve NCEA Level 3 compared to the other NCEA levels.

As with the `aggregate()` function, the `apply()` function can take a bit of getting used to, so we will now take a closer look at how the `apply()` function works.

The `apply()` function takes three main arguments.

The first argument is expected to be a matrix (or array). In the example above, we have provided the second, third, and fourth columns of the NCEA data frame, i.e., a data frame with three columns, as the first argument. This is an example of a situation that was mentioned back in Section 9.6.2, where a function will silently coerce the value that we supply to another sort of data structure if it needs to.

In this case, the `apply()` function silently coerces the data frame that we give it into a matrix. The conversion from a data frame to a matrix makes sense in this case because all three columns of the data frame that we supplied are numeric, so the data frame columns convert very naturally and predictably into a matrix (with three columns and 88 rows).

The second argument to `apply()` specifies how the array should be broken into subsets. The value `1` means split the matrix into separate rows and the value `2` means split the matrix into separate columns. In this case, we have split the matrix into columns; each column represents percentages for one NCEA level.

The third argument specifies a function to call for each subset. In this case, the `apply()` call says to take each column corresponding to an NCEA level and call the `mean()` function on each column.

The diagram below provides a conceptual view of how `apply()` works when `MARGIN=1` (apply by rows). Figure 9.11 includes a diagram that illustrates using `apply()` by columns.

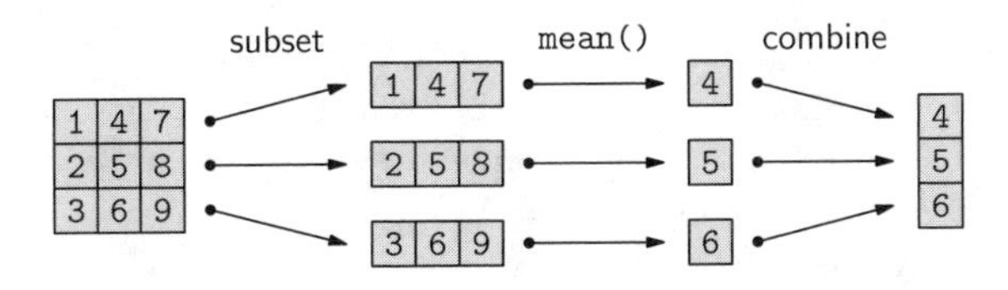

The data structure returned by `apply()` depends on how many values are returned by the function `FUN`. In the last example, we used the function `mean()`, which returns just a single value, so the overall result was a numeric vector, but if `FUN` returns more than one value, the result will be a matrix.

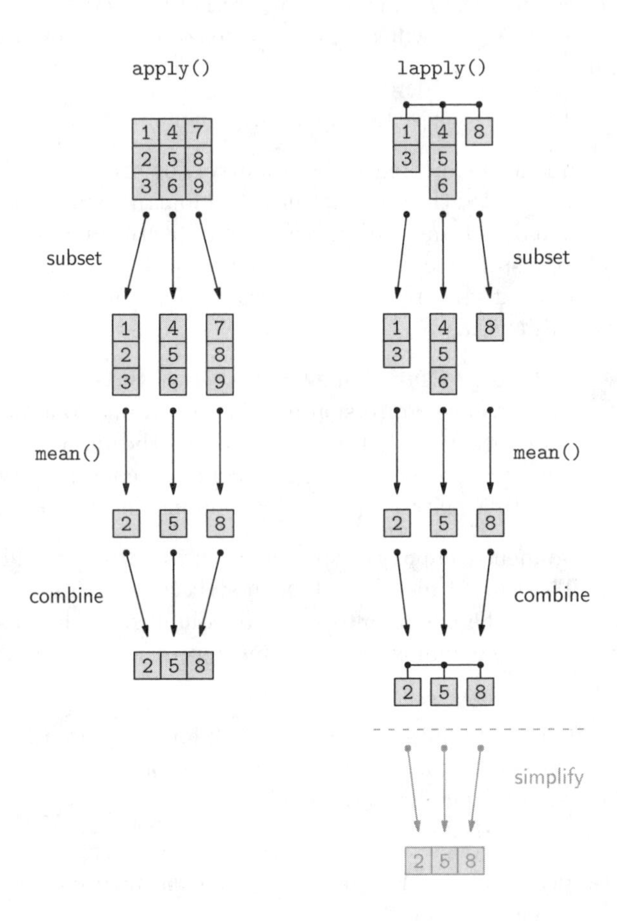

Figure 9.11: A conceptual view of how the `apply()` function (left) and the `lapply()` function (right) manipulate data by calling another function on each sub-component of a matrix or list and recombining the results into a new matrix or list. The `sapply()` function extends the `lapply()` function by performing an extra step to simplify the result to a vector (grey portion, bottom right).

For example, the following code calls `apply()` with the `range()` function to get the range of percentages at each NCEA level. The `range()` function returns two values for each NCEA level, so the result is a matrix with three columns (one for each NCEA level) and two rows—the minimum and maximum percentage at each level.

```
R> apply(NCEA[2:4], 2, range)

     Level1 Level2 Level3
[1,]    2.8    0.0    0.0
[2,]   97.4   95.7   95.7
```

The basic idea of `apply()` has similarities with the `aggregate()` and `by()` functions that we saw in the previous section. In all of these cases, we are breaking a larger data structure into smaller pieces, calling a function on each of the smaller pieces and then putting the results back together again to create a new data structure.

In the case of `aggregate()`, we start with a vector or a data frame and we end up with a data frame. The `by()` function starts with a vector or a data frame and ends up with a list-like object. With `apply()`, we start with an array and end up with a vector or a matrix.

The `lapply()` function is another example of this idea. In this case, the starting point is a *list*. The function breaks the list into its separate components, calls a function on each component, and then combines the results into a new list (see Figure 9.11).

In order to demonstrate the use of `lapply()`, we will repeat the task we just performed using `apply()`.

The following code calls `lapply()` to calculate the average percentage of students achieving each NCEA level.

```
R> lapply(NCEA[2:4], FUN=mean)

$Level1
[1] 62.26705

$Level2
[1] 61.06818

$Level3
[1] 47.97614
```

The lapply() function has only two main arguments. The first argument is a list. As with the apply() example, we have supplied a data frame—the second, third, and fourth columns of the NCEA data frame—rather than supplying a list. The lapply() function silently converts the data frame to a list, which is a natural conversion if we treat each variable within the data frame as a component of the list.

The second argument to lapply() is a function to call on each component of the list, and again we have supplied the mean() function.

The result of lapply() is a list data structure. The numeric results, the average percentages for each NCEA level, are exactly the same as they were when we used apply(); all that has changed is that these values are stored in a different sort of data structure.

Like apply(), the lapply() function comfortably handles functions that return more than a single value. For example, the following code calculates the range of percentages for each NCEA level.

```
R> lapply(NCEA[2:4], FUN=range)

$Level1
[1]  2.8 97.4

$Level2
[1]  0.0 95.7

$Level3
[1]  0.0 95.7
```

These results from lapply() are very similar to the previous results from apply(); all that has changed is that the values are stored in a different sort of data structure (a list rather than a matrix).

Whether to use apply() or lapply() will depend on what sort of data structure we are starting with and what we want to do with the result. For example, if we start with a list data structure, then lapply() is the appropriate function to use. However, if we start with a data frame, we can often use either apply() or lapply() and the choice will depend on whether we want the answer as a list or as a matrix.

The sapply() function provides a slight variation on lapply(). This function behaves just like lapply() except that it attempts to *simplify* the result by returning a simpler data structure than a list, if that is possible. For example, if we use sapply() rather than lapply() for the previous examples, we get a vector and a matrix as the result, rather than lists.

```
R> sapply(NCEA[2:4], mean)

  Level1   Level2   Level3 
62.26705 61.06818 47.97614 
```

```
R> sapply(NCEA[2:4], range)

     Level1 Level2 Level3
[1,]    2.8    0.0    0.0
[2,]   97.4   95.7   95.7
```

9.8.8 Merging

In this section, we will look at the problem of *combining* data structures.

For simple situations, the functions c(), cbind(), and rbind() are all that we need. For example, consider the two numeric vectors below.

```
R> 1:3

[1] 1 2 3
```

```
R> 4:6

[1] 4 5 6
```

The c() function is useful for combining vectors or lists to make a longer vector or a longer list.

```
R> c(1:3, 4:6)

[1] 1 2 3 4 5 6
```

The functions cbind() and rbind() can be used to combine vectors, matrices, or data frames with similar dimensions. The cbind() function creates a matrix or data frame by combining data structures side-by-side (a "column" bind) and rbind() combines data structures one above the other (a "row" bind).

```
R> cbind(1:3, 4:6)
```

```
     [,1] [,2]
[1,]    1    4
[2,]    2    5
[3,]    3    6
```

```
R> rbind(1:3, 4:6)
```

```
     [,1] [,2] [,3]
[1,]    1    2    3
[2,]    4    5    6
```

A more difficult problem arises if we want to combine two data structures that do not have the same dimensions.

In this situation, we need to first find out which rows in the two data structures correspond to each other and then combine them.

The diagram below illustrates how two data frames (left and right) can be merged to create a single data frame (middle) by matching up rows based on a common column (labeled m in the diagram).

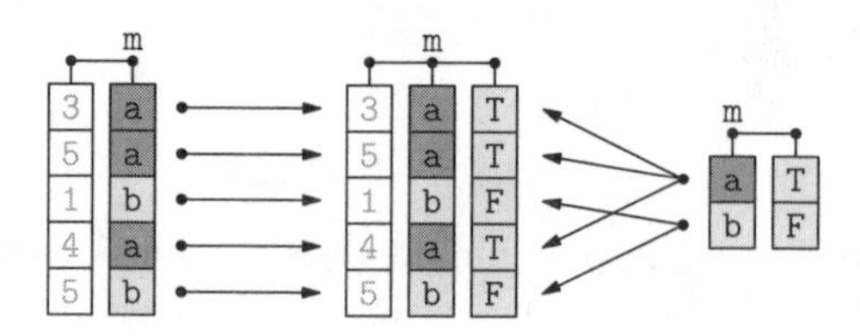

For example, the two case studies that we have been following in this section provide two data frames that contain information about New Zealand schools. The `schools` data frame contains information about the location and size of every primary and secondary school in the country, while the `NCEA` data frame contains information about the NCEA performance of every secondary school.

It would be useful to be able to combine the information from these two data frames. This would allow us to, for example, compare NCEA performance between schools of different sizes.

The `merge()` function can be used to combine data sets like this.

The following code creates a new data frame, `aucklandSchools`, that contains information from the `schools` data frame and from the `NCEA` data frame for Auckland secondary schools.

```
R> aucklandSchools <- merge(schools, NCEA,
                            by.x="Name", by.y="Name",
                            all.x=FALSE, all.y=FALSE)
```

```
R> aucklandSchools
```

```
                 Name ID      City    Auth Dec Roll Level1 Level2 Level3
35  Mahurangi College 24 Warkworth   State   8 1095   71.9   66.2   55.9
48      Orewa College 25     Orewa   State   9 1696   75.2   81.0   54.9
32   Long Bay College 27  Auckland   State  10 1611   74.5   84.2   67.2
55  Rangitoto College 28  Auckland   State  10 3022   85.0   81.7   71.6
30     Kristin School 29  Auckland Private  10 1624   93.4   27.8   36.7
19  Glenfield College 30  Auckland   State   7  972   58.4   65.5   45.6
10 Birkenhead College 31  Auckland   State   6  896   59.8   65.7   50.4
...
```

The first two arguments to the `merge()` function are the data frames to be merged, in this case, `schools` and `NCEA`.

The `by.x` and `by.y` arguments specify the names of the variables to use to match the rows from one data frame with the other. In this case, we use the `Name` variable from each data frame to determine which rows belong with each other.

The `all.x` argument says that the result should only include schools from the `schools` data frame that have the same name as a school in the `NCEA` data frame. The `all.y` argument means that a school in `NCEA` must have a match in `schools` or it is not included in the final result.

By default, `merge()` will match on any variable names that the two data frames have in common, and it will only include rows in the final result if there is a match between the data frames. In this case, we could have just used the following expression to get the same result:

```
R> merge(schools, NCEA)
```

9.8.9 Flashback: Database joins

In the terminology of Section 7.2.4, the `merge()` function is analogous to a **database join**.

In relational database terms, `by.x` and `by.y` correspond to a primary-key/foreign-key pairing. The `all.x` and `all.y` arguments control whether the join is an inner join or an outer join.

The `merge()` call above roughly corresponds to the following SQL query.

```
SELECT *
    FROM schools INNER JOIN NCEA
        ON schools.Name = NCEA.Name;
```

9.8.10 Splitting

Instead of combining two data frames, it is sometimes useful to break a data frame into several pieces. For example, we might want to perform a separate analysis on the New Zealand schools for each different type of school ownership.

The `split()` function can be used to break up a data frame into smaller data frames, based on the levels of a factor. The result is a list, with one component for each level of the factor.

The diagram below illustrates how the `split()` function uses the values within a factor to break a data frame (left) into a list containing two separate data frames (right).

The result is effectively several subsets of the original data frame, but with the subsets all conveniently contained within a single overall data structure (a list), ready for subsequent processing or analysis.

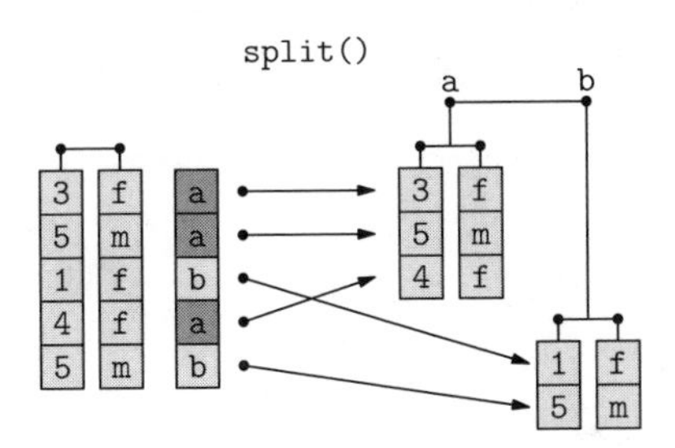

In the following code, the `Rolls` variable from the `schools` data frame is broken into four pieces based on the `Auth` variable.

```
R> schoolRollsByAuth <- split(schools$Roll, schools$Auth)
```

The resulting list, `schoolRollsByAuth`, has four components, each named after one level of the `Auth` variable and each containing the number of students enrolled at schools of the appropriate type. The following code uses the `str()` function that we saw in Section 9.6.7 to show a brief view of the basic structure of this list.

```
R> str(schoolRollsByAuth)
```

```
List of 4
 $ Other           : int 51
 $ Private         : int [1:99] 255 39 154 73 83 25 95 85 ..
 $ State           : int [1:2144] 318 200 455 86 577 329 6..
 $ State Integrated: int [1:327] 438 26 191 560 151 114 12..
```

Because the result of `split()` is a list, it is often used in conjunction with `lapply()` to perform a task on each component of the resulting list. For example, we can calculate the average roll for each type of school ownership with the following code.

```
R> lapply(schoolRollsByAuth, mean)

$Other
[1] 51

$Private
[1] 308.798

$State
[1] 300.6301

$`State Integrated`
[1] 258.3792
```

This result can be compared to the result we got using the `aggregate()` function on page 298.

9.8.11 Reshaping

For multivariate data sets, where a single individual or case is measured several times, there are two common formats for storing the data.

The so-called "wide" format uses a separate variable for each measurement, and the "long" format for a data set has a *single observation* on each row.

The diagram below illustrates the difference between wide format (on the left), where each row contains multiple measurements (`t1`, `t2`, and `t3`) for a single individual, and long format (on the right), where each row contains a single measurement and there are multiple rows for each individual.

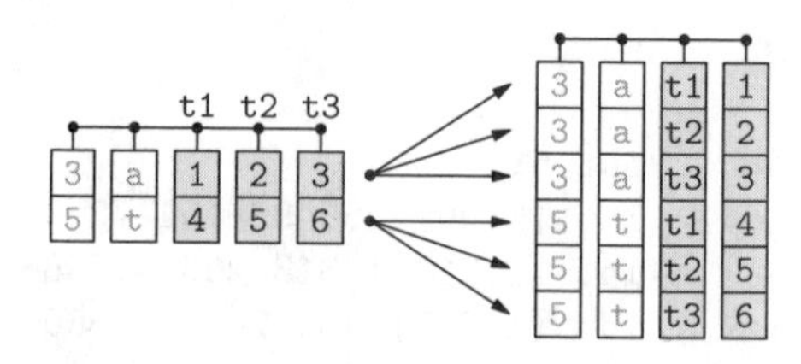

For some data processing tasks, it is useful to have the data in wide format, but for other tasks, the long format is more convenient, so it is useful to be able to convert a data set from one format to the other.

The NCEA data frame that we have been using is in wide format because it has a row for each school and a column for each NCEA level. There is one row per school and *three observations* on each row.

```
R> NCEA
```

```
                                   Name Level1 Level2 Level3
1                     Al-Madinah School   61.5   75.0    0.0
2                     Alfriston College   53.9   44.1    0.0
3 Ambury Park Centre for Riding Therapy   33.3   20.0    0.0
4                        Aorere College   39.5   50.2   30.6
5        Auckland Girls' Grammar School   71.2   78.9   55.5
6                      Auckland Grammar   22.1   30.8   26.3
7    Auckland Seventh-Day Adventist H S   50.8   34.8   48.9
...
```

In the long format of the NCEA data set, each school would have three rows, one for each NCEA level. What this format looks like is shown below. We will go on to show *how* to obtain this format.

```
                                   Name variable value
1                     Alfriston College   Level1  53.9
2                     Alfriston College   Level2  44.1
3                     Alfriston College   Level3   0.0
4                     Al-Madinah School   Level1  61.5
5                     Al-Madinah School   Level2  75.0
6                     Al-Madinah School   Level3   0.0
7 Ambury Park Centre for Riding Therapy   Level1  33.3
...
```

There are several ways to perform the transformation between wide and long formats in R, but we will focus in this section on the **reshape** package because it provides a wider range of options and a more intuitive set of arguments.

```
R> library("reshape")
```

The two main functions in the **reshape** package are called `melt()` and `cast()`. The `melt()` function is used to convert a data frame into long format, and the `cast()` function can then be used to reshape the data into a variety of other forms.

Melting

The following code creates the long format version of the `NCEA` data frame using the `melt()` function.

```
R> longNCEA <- melt(NCEA, measure.var=2:4)
```

The `melt()` function typically requires two arguments. The first argument provides the data frame to melt. The second argument is *either* `measure.var` to specify which columns of the data frame are observations (**reshape** calls them "measured" variables) *or* `id.var` to specify which columns of the data frame are just values that identify the individual or case (**reshape** calls these "id" variables).

In the example above, we have specified that the second, third, and fourth columns are observations. The `melt()` function infers that the first column is therefore an "id" variable. The following code produces an identical result, using the `id.var` argument instead. In this case, we have specified that the `Name` variable is an "id" variable and the `merge()` function will treat all other columns as "measured" variables. This code also demonstrates that columns in the data frame can be specified by name as well as by number.

```
R> longNCEA <- melt(NCEA, id.var="Name")
```

```
                                 Name variable value
1                   Alfriston College   Level1  53.9
2                   Alfriston College   Level2  44.1
3                   Alfriston College   Level3   0.0
4                   Al-Madinah School   Level1  61.5
5                   Al-Madinah School   Level2  75.0
6                   Al-Madinah School   Level3   0.0
7 Ambury Park Centre for Riding Therapy   Level1  33.3
...
```

The `longNCEA` data frame has 264 rows because each of the original 88 schools now occupies three rows.

```
R> dim(longNCEA)
```

```
[1] 264   3
```

The original four columns in NCEA have become three columns in longNCEA. The first column in longNCEA is the same as the first column of NCEA, except that all of the values repeat three times.

The second column in longNCEA is called variable and it records which column of NCEA that each row of longNCEA has come from. For example, all values from the original Level1 column in NCEA have the value Level1 in the new variable column in longNCEA.

The third column in longNCEA is called value and this contains all of the data from the Level1, Level2, and Level3 columns of NCEA.

Casting

We now turn our attention to the other main function in the **reshape** package, the cast() function.

As a simple demonstration of the cast() function, we will use it to recreate the original wide format of the NCEA data set. The following code performs this transformation.

```
R> cast(longNCEA, Name ~ variable)
```

```
                                  Name Level1 Level2 Level3
1                   Alfriston College   53.9   44.1    0.0
2                   Al-Madinah School   61.5   75.0    0.0
3 Ambury Park Centre for Riding Therapy 33.3   20.0    0.0
4                     Aorere College   39.5   50.2   30.6
5        Auckland Girls' Grammar School 71.2   78.9   55.5
6                    Auckland Grammar   22.1   30.8   26.3
7    Auckland Seventh-Day Adventist H S 50.8   34.8   48.9
...
```

The first argument to cast() is a data frame in long format.

The second argument to cast() is a **formula**.

Variables on the left-hand side of this formula are used to form the rows of the result and variables on the right-hand side are used to form columns. In the above example, the result contains a row for each different school, based on the Name of the school, and a column for each different NCEA level, where the variable column specifies the NCEA level.

The following code uses a different formula to demonstrate the power that `cast()` provides for reshaping data. In this case, the data set has been transposed so that each *column* is now a different school and each *row* corresponds to one of the NCEA levels.

```
R> tNCEA <- cast(longNCEA, variable ~ Name)
```

```
R> tNCEA
```

```
  variable Alfriston College Al-Madinah School ...
1   Level1              53.9              61.5 ...
2   Level2              44.1              75.0 ...
3   Level3               0.0               0.0 ...
```

This data structure has 3 rows and 89 columns: one row for each NCEA level and one column for each school, plus an extra column containing the names of the NCEA levels.

```
R> dim(tNCEA)
```

```
[1]  3 89
```

Now that we have seen a number of techniques for manipulating data sets, the next section presents a larger case study that combines several of these techniques to perform a more complex exploration of a data set.

9.8.12 Case study: Utilities

A compact fluorescent light bulb. This sort of light bulb lasts up to 16 times longer and consumes about one quarter of the power of a comparable incandescent light bulb.

A resident of Baltimore, Maryland, in the United States collected data from his residential gas and electricity power bills over 8 years. The data are in

```
start    therms  gas KWHs         elect    temp    days
10-Jun-98        9       16.84    613      63.80   75     40
20-Jul-98        6       15.29    721      74.21   76     29
18-Aug-98        7       15.73    597      62.22   76     29
16-Sep-98        42      35.81    460      43.98   70     33
19-Oct-98        105     77.28    314      31.45   57     29
17-Nov-98        106     77.01    342      33.86   48     30
17-Dec-98        200     136.66   298      30.08   40     33
19-Jan-99        144     107.28   278      28.37   37     30
18-Feb-99        179     122.80   253      26.21   39     29
 ...
```

Figure 9.12: The first few lines of the Utilities data set in a plain text, white-space-delimited format.

a text file called **baltimore.txt** and include the start date for the bill, the number of therms of gas used and the amount charged, the number of kilowatt hours of electricity used and the amount charged, the average daily outdoor temperature (as reported on the bill), and the number of days in the billing period. Figure 9.12 shows the first few lines of the data file.

Several events of interest occurred in the household over the time period during which these values were recorded, and one question of interest was to determine whether any of these events had any effect on the energy consumption of the household.

The events were:

- An additional resident moved in on July 31st 1999.
- Two storm windows were replaced on April 22nd 2004.
- Four storm windows were replaced on September 1st 2004.
- An additional resident moved in on December 18th 2005.

The text file containing the energy usage can be read conveniently using the `read.table()` function, with the `header=TRUE` argument specified to use the variable names on the first line of the file. We also use `as.is=TRUE` to keep the dates as text for now.

```
R> utilities <- read.table(file.path("Utilities",
                                     "baltimore.txt"),
                           header=TRUE, as.is=TRUE)
```

```
R> utilities
```

```
      start therms   gas KWHs elect temp days
1 10-Jun-98      9 16.84  613 63.80   75   40
2 20-Jul-98      6 15.29  721 74.21   76   29
3 18-Aug-98      7 15.73  597 62.22   76   29
4 16-Sep-98     42 35.81  460 43.98   70   33
5 19-Oct-98    105 77.28  314 31.45   57   29
6 17-Nov-98    106 77.01  342 33.86   48   30
...
```

The first thing we want to do is to convert the first variable into actual dates. This will allow us to perform calculations and comparisons on the date values.

```
R> utilities$start <- as.Date(utilities$start,
                              "%d-%b-%y")
```

```
R> utilities
```

```
       start therms   gas KWHs elect temp days
1 1998-06-10      9 16.84  613 63.80   75   40
2 1998-07-20      6 15.29  721 74.21   76   29
3 1998-08-18      7 15.73  597 62.22   76   29
4 1998-09-16     42 35.81  460 43.98   70   33
5 1998-10-19    105 77.28  314 31.45   57   29
6 1998-11-17    106 77.01  342 33.86   48   30
...
```

The next thing that we will do is break the data set into five different time "phases," using the four significant events as breakpoints; we will be interested in the average daily charges for each of these phases.

To keep things simple, we will just determine the phase based on whether the billing period *began* before a significant event.

The six critical dates that we will use to categorize each billing period are the start of the first billing period, the dates at which the significant events occurred, and the start of the last billing period.

```
R> breakpoints <- c(min(utilities$start),
                    as.Date(c("1999-07-31", "2004-04-22",
                              "2004-09-01", "2005-12-18")),
                    max(utilities$start))
R> breakpoints
```

```
[1] "1998-06-10" "1999-07-31" "2004-04-22" "2004-09-01"
[5] "2005-12-18" "2006-08-17"
```

We can use the `cut()` function to convert the `start` variable, which contains the billing period start dates, into a `phase` variable, which is a factor.

```
R> phase <- cut(utilities$start, breakpoints,
                include.lowest=TRUE, labels=FALSE)
R> phase
```

```
 [1] 1 1 1 1 1 1 1 1 1 1 1 1 1 2 2 2 2 2 2 2 2 2 2 2 2 2 2 2
[29] 2 2 2 2 2 2 2 2 2 2 2 2 2 2 2 2 2 2 2 2 2 2 2 2 2 2 2 2
[57] 2 2 2 2 2 2 2 2 2 2 2 2 3 3 3 3 4 4 4 4 4 5 4 4 4 4 4 4
[85] 4 4 4 4 4 5 5 5 5 5 5 5
```

Each billing period now has a corresponding phase.

One important point to notice about these `phase` values is that they are *not* strictly increasing. In particular, the 78th value is a `5` amongst a run of `4`s. This reveals that the billing periods in the original file were *not* entered in strictly chronological order.

Now that we have each billing period assigned to a corresponding phase, we can sum the energy usage and costs for each phase. This is an application of the `aggregate()` function. The following code subsets the `utilities` data frame to remove the first column, the billing period start dates, and then sums the values in all of the remaining columns for each different phase.

```
R> phaseSums <- aggregate(utilities[c("therms", "gas", "KWHs",
                                      "elect", "days")],
                          list(phase=phase), sum)
```

```
R> phaseSums

  phase therms     gas  KWHs   elect days
1     1    875  703.27  5883  602.84  406
2     2   5682 5364.23 24173 2269.14 1671
3     3     28   76.89  1605  170.27  124
4     4   1737 2350.63  5872  567.91  483
5     5    577  847.38  3608  448.61  245
```

We can now divide the totals for each phase by the number of days in each phase to obtain an average usage and cost per day for each phase.

```
R> phaseAvgs <- phaseSums[2:5]/phaseSums$days
```

For inspecting these values, it helps if we round the values to two significant figures.

```
R> signif(phaseAvgs, 2)

  therms  gas KWHs elect
1   2.20 1.70   14   1.5
2   3.40 3.20   14   1.4
3   0.23 0.62   13   1.4
4   3.60 4.90   12   1.2
5   2.40 3.50   15   1.8
```

The division step above works because when a data frame is divided by a vector, each variable in the data frame gets divided by the vector. This is not necessarily obvious from the code; a more explicit way to perform the operation is to use the `sweep()` function, which forces us to explicitly state that, for each row of the data frame (`MARGIN=1`), we are dividing (`FUN="/"`) by the corresponding value from a vector (`STAT=phase$days`).

```
R> phaseSweep <- sweep(phaseSums[2:5],
                       MARGIN=1, STAT=phaseSums$days, FUN="/")
```

Looking at the average daily energy values for each phase, the values that stand out are the gas usage and cost during phase 3 (after the first two storm windows were replaced, but before the second set of four storm windows were replaced). The naive interpretation is that the first two storm windows were incredibly effective, but the next four storm windows actually made things worse again!

At first sight this appears strange, but it is easily explained by the fact that phase 3 coincided with summer months, as shown below using the `table()` function. This code produces a rough count of how many times each month occurred in each phase. The function `months()` extracts the names of the months from the dates in the `start` variable; each billing period is assigned to a particular month based on the month that the period *started* in (which is why this is a *rough* count).

```
R> table(months(utilities$start), phase)[month.name, ]
```

```
           phase
            1 2 3 4 5
  January   1 5 0 1 1
  February  1 5 0 1 1
  March     0 5 0 1 1
  April     1 5 0 1 1
  May       1 3 1 1 1
  June      2 4 1 1 1
  July      2 4 1 1 1
  August    1 5 1 1 1
  September 1 5 0 2 0
  October   1 5 0 2 0
  November  1 4 0 2 0
  December  1 5 0 2 0
```

Subsetting the resulting table by the predefined symbol `month.name` just rearranges the order of the rows of the table; this ensures that the months are reported in calendar order rather than alphabetical order.

We would expect the gas usage (for heating) to be a lot lower during summer. This is confirmed by (roughly) calculating the daily average usage and cost for each month.

Again we must take care to get the answer in calendar order; this time we do that by creating a factor that is based on converting the start of each billing period into a month, with the order of the levels of this factor explicitly set to be `month.name`.

```
R> billingMonths <- factor(months(utilities$start),
                           levels=month.name)
```

We then use aggregate() to sum the usage, costs, and days for the billing periods, grouped by this billingMonths factor.

```
R> months <-
       aggregate(utilities[c("therms", "gas", "KWHs",
                             "elect", "days")],
                 list(month=billingMonths),
                 sum)
```

Finally, we divide the usage and cost totals by the number of days in the billing periods. So that we can easily comprehend the results, we cbind() the month names on to the averages and round the averages using signif().

```
R> cbind(month=months$month,
         signif(months[c("therms", "gas",
                         "KWHs", "elect")]/months$days,
                2))
```

```
       month therms  gas KWHs elect
1    January   7.80 7.50   13   1.2
2   February   5.90 5.90   11   1.0
3      March   3.90 4.00   12   1.1
4      April   1.70 1.90   11   1.0
5        May   0.65 0.99   12   1.3
6       June   0.22 0.57   16   1.8
7       July   0.21 0.59   22   2.3
8     August   0.23 0.60   17   1.8
9  September   1.20 1.30   14   1.3
10   October   3.40 3.50   12   1.1
11  November   5.40 5.50   14   1.2
12  December   7.20 7.10   13   1.2
```

These results are presented graphically in Figure 9.13.

This example has demonstrated a number of data manipulation techniques in a more realistic setting. The simple averages that we have calculated serve to show that any attempt to determine whether the significant events led to a significant change in energy consumption and cost for this household is clearly going to require careful analysis.

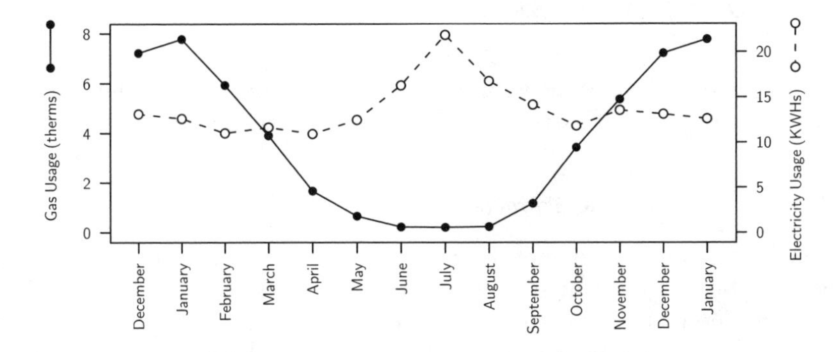

Figure 9.13: A plot of the average daily energy usage, by month, for the Utilities data set. Gas usage is represented by the solid line and electricity usage is represented by the dashed line.

Recap

Simple manipulations of data include sorting values, binning numeric variables, and producing tables of counts for categorical variables.

Aggregation involves producing a summary value for each group in a data set.

An "apply" operation calculates a summary value for each variable in a data frame, or for each row or column of a matrix, or for each component of a list.

Data frames can be merged just like tables can be joined in a database. They can also be split apart.

Reshaping involves transforming a data set from a row-per-subject, "wide", format to a row-per-observation, "long", format (among other things).

9.9 Text processing

The previous section dealt with techniques that are useful for for manipulating a variety of data structures and a variety of data types.

This section is focused solely on data manipulation techniques for working with *vectors* containing *character values.*

Plain text is a very common format for storing information, so it is very useful to be able to manipulate text. It may be necessary to convert a data set from one text format to another. It is also common to search for and extract important keywords or specific patterns of characters from within a large set of text.

We will introduce some of the basic text processing ideas via a simple example.

9.9.1 Case study: The longest placename

One of the longest placenames in the world, with a total of 85 characters, is the Maori name for a hill in the Hawkes Bay region, on the east coast of the North Island of New Zealand.

One of the longest placenames in the world is attributed to a hill in the Hawke's Bay region of New Zealand. The name (in Maori) is ...

```
Taumatawhakatangihangakoauauotamateaturipukakapikimaungahoronukupokaiwhenuakitanatahu
```

... which means "The hilltop where Tamatea with big knees, conqueror of mountains, eater of land, traveler over land and sea, played his koauau [flute] to his beloved."

Children at an Auckland primary school were given a homework assignment that included counting the number of letters in this name. This task of counting the number of characters in a piece of text is a simple example of what we will call **text processing** and is the sort of task that often comes up when working with data that have been stored in a text format.

Counting the number of characters in a piece of text is something that any programming language will do. Assuming that the name has been saved into a text file called placename.txt, here is how to use the scan() function to read the name into R, as a character vector of length 1.

```
R> placename <- scan(file.path("Placename", "placename.txt"),
                     "character")
```

The first argument provides the name and location of the file and the second argument specifies what sort of data type is in the file. In this case, we are reading a single character value from a file.

We can now use the nchar() function to count the number of characters in this text.

```
R> nchar(placename)

[1] 85
```

Counting characters is a very simple text processing task, though even with something that simple, performing the task using a computer is much more likely to get the right answer. We will now look at some more complex text processing tasks.

The homework assignment went on to say that, in Maori, the combinations 'ng' and 'wh' can be treated as a single letter. Given this, how many letters are in the placename? There are two possible approaches: convert every 'ng' and 'wh' to a single letter and recount the number of letters, or count the number of 'ng's and 'wh's and subtract that from the total number of characters. We will consider both approaches because they illustrate two different text processing tasks.

For the first approach, we could try counting all of the 'ng's and 'wh's as single letters by searching through the text and converting all of them into single characters and then redoing the count. In R, we can perform this search-and-replace task using the gsub() function, which takes three arguments: a pattern to search for, a replacement value, and the text to search within. The result is the original text with the pattern replaced. Because we are only counting letters, it does not matter which letter we choose as a replacement. First, we replace occurrences of 'ng' with an underscore character.

```
R> replacengs <- gsub("ng", "_", placename)
R> replacengs
```

```
[1] "Taumatawhakata_iha_akoauauotamateaturipukakapikimau_ahoronukupokaiwhenuakitanatahu"
```

Next, we replace the occurrences of 'wh' with an underscore character.

```
R> replacewhs <- gsub("wh", "_", replacengs)
R> replacewhs
```

```
[1] "Taumata_akata_iha_akoauauotamateaturipukakapikimau_ahoronukupokai_enuakitanatahu"
```

Finally, we count the number of letters in the resulting text.

```
R> nchar(replacewhs)

[1] 80
```

The alternative approach involves just finding out how many 'ng's and 'wh's are in the text and subtracting that number from the original count. This simple step of searching within text for a pattern is yet another common text processing task. There are several R functions that perform variations on this task, but for this example we need the function gregexpr() because it returns *all* of the matches within a piece of text. This function takes two arguments: a pattern to search for and the text to search within. The result is a vector of the starting positions of the pattern within the text, with an attribute that gives the lengths of each match. Figure 9.14 includes a diagram that illustrates how this function works.

```
R> ngmatches <- gregexpr("ng", placename)[[1]]
R> ngmatches

[1] 15 20 54
attr(,"match.length")
[1] 2 2 2
```

This result shows that the pattern 'ng' occurs three times in the placename, starting at character positions 15, 20, and 54, respectively, and that the

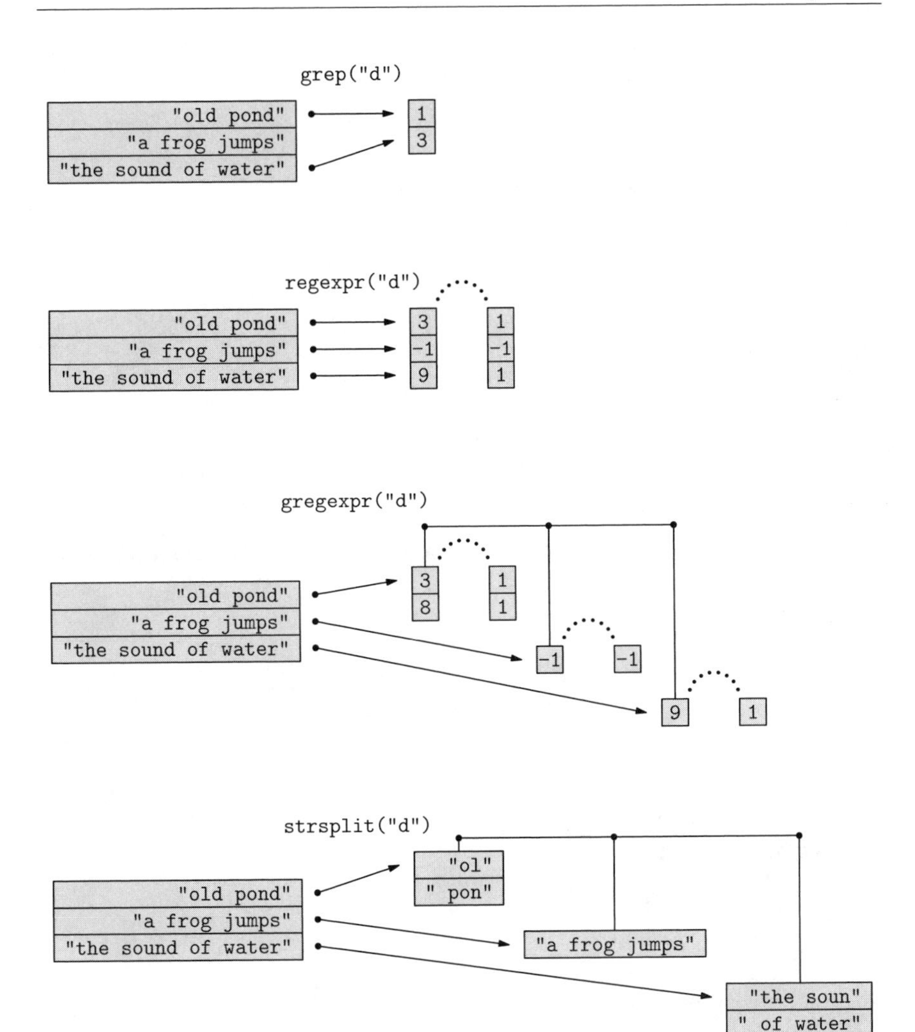

Figure 9.14: A conceptual view of how various text processing functions work. In each case, the text is a character vector with three elements (an English translation of a famous haiku by Matsuo Bashō). From top to bottom, the diagrams show: searching for the letter "d" using `grep()`, `regexpr()`, and `gregexpr()`; and splitting the text on the letter "d" using `strsplit()`. The result of `grep()` is a numeric vector indicating which elements match. The result of `regexpr()` is a numeric vector indicating the position of the *first* match in each element, *plus* an attribute indicating the length of each match. The result of `gregexpr()` is a *list* of numeric vectors with attributes similar to `regexpr()` but giving *all* matches in each element. The result of `strsplit()` is a list of character vectors, where each component is the result of splitting one element of the original text.

length of the match is 2 characters in each case. Here is the result of searching for occurrences of 'wh':

```
R> whmatches <- gregexpr("wh", placename)[[1]]
R> whmatches
```

```
[1]  8 70
attr(,"match.length")
[1] 2 2
```

The return value of `gregexpr()` is a list to allow for more than one piece of text to be searched at once. In this case, we are only searching a single piece of text, so we just need the first component of the result.

We can use the `length()` function to count how many matches there were in the text.

```
R> length(ngmatches)
```

```
[1] 3
```

```
R> length(whmatches)
```

```
[1] 2
```

The final answer is simple arithmetic.

```
R> nchar(placename) -
     (length(ngmatches) + length(whmatches))
```

```
[1] 80
```

For the final question in the homework assignment, the students had to count how many times each letter appeared in the placename (treating 'wh' and 'ng' as separate letters again).

One way to do this in R is by breaking the placename into individual characters and creating a table of counts. Once again, we have a standard text processing task: breaking a single piece of text into multiple pieces. The strsplit() function performs this task in R. It takes two arguments: the text to break up and a pattern which is used to decide where to split the text. If we give the value NULL as the second argument, the text is split at each character. Figure 9.14 includes a diagram that illustrates how this function works.

```
R> nameLetters <- strsplit(placename, NULL)[[1]]
R> nameLetters

 [1] "T" "a" "u" "m" "a" "t" "a" "w" "h" "a" "k" "a" "t" "a"
[15] "n" "g" "i" "h" "a" "n" "g" "a" "k" "o" "a" "u" "a" "u"
[29] "o" "t" "a" "m" "a" "t" "e" "a" "t" "u" "r" "i" "p" "u"
[43] "k" "a" "k" "a" "p" "i" "k" "i" "m" "a" "u" "n" "g" "a"
[57] "h" "o" "r" "o" "n" "u" "k" "u" "p" "o" "k" "a" "i" "w"
[71] "h" "e" "n" "u" "a" "k" "i" "t" "a" "n" "a" "t" "a" "h"
[85] "u"
```

Again, the result is a list to allow for breaking up multiple pieces of text at once. In this case, because we only have one piece of text, we are only interested in the first component of the list.

One minor complication is that we want the uppercase 'T' to be counted as a lowercase 't'. The function tolower() performs this task.

```
R> lowerNameLetters <- tolower(nameLetters)
R> lowerNameLetters

 [1] "t" "a" "u" "m" "a" "t" "a" "w" "h" "a" "k" "a" "t" "a"
[15] "n" "g" "i" "h" "a" "n" "g" "a" "k" "o" "a" "u" "a" "u"
[29] "o" "t" "a" "m" "a" "t" "e" "a" "t" "u" "r" "i" "p" "u"
[43] "k" "a" "k" "a" "p" "i" "k" "i" "m" "a" "u" "n" "g" "a"
[57] "h" "o" "r" "o" "n" "u" "k" "u" "p" "o" "k" "a" "i" "w"
[71] "h" "e" "n" "u" "a" "k" "i" "t" "a" "n" "a" "t" "a" "h"
[85] "u"
```

Now it is a simple matter of calling the tablefunction to produce a table of counts of the letters.

```
R> letterCounts <- table(lowerNameLetters)
R> letterCounts

lowerNameLetters
 a  e  g  h  i  k  m  n  o  p  r  t  u  w
22  2  3  5  6  8  3  6  5  3  2  8 10  2
```

As well as pulling text apart into smaller pieces as we have done so far, we also need to be able to put several pieces of text together to make a single larger piece of text.

For example, if we begin with the individual letters of the placename, as in the character vector `nameLetters`, how do we combine the letters to make a single character value? In R, this can done with the `paste()` function.

The `paste()` function can be used to combine separate character vectors *or* to combine the character values within a single character vector. In this case, we want to perform the latter task.

We have a character vector containing 85 separate character values.

```
R> nameLetters

 [1] "T" "a" "u" "m" "a" "t" "a" "w" "h" "a" "k" "a" "t" "a"
[15] "n" "g" "i" "h" "a" "n" "g" "a" "k" "o" "a" "u" "a" "u"
[29] "o" "t" "a" "m" "a" "t" "e" "a" "t" "u" "r" "i" "p" "u"
[43] "k" "a" "k" "a" "p" "i" "k" "i" "m" "a" "u" "n" "g" "a"
[57] "h" "o" "r" "o" "n" "u" "k" "u" "p" "o" "k" "a" "i" "w"
[71] "h" "e" "n" "u" "a" "k" "i" "t" "a" "n" "a" "t" "a" "h"
[85] "u"
```

The following code combines the individual character values to make the complete placename. The `collapse` argument specifies that the character vector should be collapsed into a single character value with, in this case (`collapse=""`), nothing in between each character.

```
R> paste(nameLetters, collapse="")

[1] "Taumatawhakatangihangakoauauotamateaturipukakapikimaungahoronukupokaiwhenuakitanatahu"
```

This section has introduced a number of functions for counting letters in text, transforming text, breaking text apart, and putting it back together again. More examples of the use of these functions are given in the next section and in case studies later on.

9.9.2 Regular expressions

Two of the tasks we looked at when working with the long Maori placename in the previous case study involved treating both 'ng' and 'wh' as if they were single letters by replacing them both with underscores. We performed the task in two steps: convert all occurrences of 'ng' to an underscore, and then convert all occurrences of 'wh' to an underscore. Conceptually, it would be simpler, and more efficient, to perform the task in a single step: convert all occurrences of 'ng' *and* 'wh' to underscore characters. Regular expressions allow us to do this.

With the placename in the variable called `placename`, converting both 'ng' and 'wh' to underscores in a single step is achieved as follows:

```
R> gsub("ng|wh", "_", placename)
```

```
[1] "Taumata_akata_iha_akoauauotamateaturipukakapikimau_ahoronukupokai_enuakitanatahu"
```

The regular expression we are using, `ng|wh`, describes a **pattern**: the character 'n' followed by the character 'g' *or* the character 'w' followed by the character 'h'. The vertical bar, `|`, is a **metacharacter**. It does not have its normal meaning, but instead denotes an optional pattern; a match will occur if the text contains either the pattern to the left of the vertical bar or the pattern to the right of the vertical bar. The characters 'n', 'g', 'w', and 'h' are all **literals**; they have their normal meaning.

A regular expression consists of a mixture of **literal** characters, which have their normal meaning, and **metacharacters**, which have a special meaning. The combination describes a **pattern** that can be used to find matches amongst text values.

A regular expression may be as simple as a literal word, such as `cat`, but regular expressions can also be quite complex and express sophisticated ideas, such as `[a-z]{3,4}[0-9]{3}`, which describes a pattern consisting of either three or four lowercase letters followed by any three digits.

Just like all of the other technologies in this book, there are several different versions of regular expressions, however, rather than being numbered, the different versions of regular expressions have different names: there are

Basic regular expressions, Extended (POSIX) regular expressions, and Perl-Compatible regular expressions (PCRE). We will assume Extended regular expressions in this book, though for the level of detail that we encounter, the differences between Extended regular expressions and PCRE are not very important. Basic regular expressions, as the name suggests, have fewer features than the other versions.

There are two important components required to use regular expressions: we need to be able to write regular expressions *and* we need software that understands regular expressions in order to search text for the pattern specified by the regular expression.

We will focus on the use of regular expressions with R in this book, but there are many other software packages that understand regular expressions, so it should be possible to use the information in this chapter to write effective regular expressions within other software environments as well. One caveat is that R consumes backslashes in text, so it is necessary to type a double backslash in order to obtain a single backslash in a regular expression. This means that we often avoid the use of backslashes in regular expressions in R and, when we are forced to use backslashes, the regular expressions will look more complicated than they would in another software setting.

Something to keep in mind when writing regular expressions (especially when trying to determine why a regular expression is not working) is that most software that understands regular expressions will perform "eager" and "greedy" searches. This means that the searches begin from the start of a piece of text, the first possible match succeeds (even if a "better" match might exist later in the text), and as many characters as possible are matched. In other words, the first and longest possible match is found by each component of a regular expression. A common problem is to have a later component of a regular expression fail to match because an earlier component of the expression has consumed the entire piece of text all by itself.

For this reason, it is important to remember that regular expressions are a small computer language of their own and should be developed with just as much discipline and care as we apply to writing any computer code. In particular, a complex regular expression should be built up in smaller pieces in order to understand how each component of the regular expression works before adding further components.

The next case study looks at some more complex uses and provides some more examples. Chapter 11 describes several other important metacharacters that can be used to build more complex regular expressions.

9.9.3 Case study: Rusty wheat

Cereal crops account for almost half of global food production. Maize, rice, and wheat make up almost 90% of that production, with barley (pictured) fourth on the list.

As part of a series of field trials conducted by the Institut du Végétal in France,[3] data were gathered on the effect of the disease *Septoria tritici* on wheat. The amount of disease on individual plants was recorded using data collection forms that were filled in by hand by researchers in the field.

In 2007, due to unusual climatic conditions, two other diseases, *Puccinia recondita* ("brown rust") and *Puccinia striiformis* ("yellow rust") were also observed to be quite prevalent. The data collection forms had no specific field for recording the amount of rust on each wheat plant, so data were recorded ad hoc in a general area for "diverse observations".

In this case study, we will not be interested in the data on *Septoria tritici.* Those data were entered into a standard form using established protocols, so the data were relatively tidy.

Instead, we will focus on the yellow rust and brown rust data because these data were not anticipated, so the data were recorded quite messily. The data included other comments unrelated to rust and there were variations in how the rust data were expressed by different researchers.

This lack of structure means that the rust data cannot be read into R using the functions that expect a regular format, such as `read.table()` and `read.fwf()` (see Section 9.7.3). This provides us with an example where text processing tools allow us to work with data that have an irregular structure.

The yellow and brown rust data were transcribed verbatim into a plain text file, as shown in Figure 9.15.

Fortunately, for the purposes of recovering these results, some basic features of the data are consistent.

[3]Thanks to David Gouache, Arvalis—Institut du Végétal.

```
lema, rb 2%
rb 2%
rb 3%
rb 4%
rb 3%
rb 2%,mineuse
rb
rb
rb 12
rb
rj 30%
rb
rb
rb 25%
rb
rb
rb
rj 10, rb 4
```

Figure 9.15: Data recording the occurrence of brown or yellow rust diseases on wheat plants, in a largely unstructured plain text format. Each line represents one wheat plant. The text `rb` followed by a number represents an amount of brown rust and the text `rj` followed by a number represents an amount of yellow rust.

Each line of data represents one wheat plant. If brown rust was present, the line contains the letters `rb`, followed by a space, followed by a number indicating the percentage of the plant affected by the rust (possibly with a percentage sign). If the plant was afflicted by yellow rust, the same pattern applies except that the letters `rj` are used. It is possible for both diseases to be present on the same plant (see the last line of data in Figure 9.15).

The abbreviations `rb` and `rj` were used because the French common names for the diseases are *rouille brune* and *rouille jaune*.

For this small set of recordings, the data could be extracted by hand. However, the full data set contains many more records so we will develop a code solution that uses regular expressions to recover the rust data from these recordings.

The first step is to get the data into R. We can do this using the `readLines()` function, which will create a character vector with one element for each line of recordings.

```
R> wheat <- readLines(file.path("Wheat", "wheat.txt"))
R> wheat

 [1] "lema, rb 2%"  "rb 2%"         "rb 3%"
 [4] "rb 4%"        "rb 3%"         "rb 2%,mineuse"
 [7] "rb"           "rb"            "rb 12"
[10] "rb"           "rj 30%"        "rb"
[13] "rb"           "rb 25%"        "rb"
[16] "rb"           "rb"            "rj 10, rb 4"
```

What we want to end up with are two variables, one recording the amount of brown rust on each plant and one recording the amount of yellow rust.

Starting with brown rust, the first thing we could do is find out which plants have any brown rust on them. The following code does this using the `grep()` function. The result is a vector of indices that tells us which lines contain the pattern we are searching for. Figure 9.14 includes a diagram that illustrates how this function works.

```
R> rbLines <- grep("rb [0-9]+", wheat)
R> rbLines

[1]  1  2  3  4  5  6  9 14 18
```

The regular expression in this call demonstrates two more important examples of metacharacters. The square brackets, '`[`' and '`]`', are used to describe a **character set** that will be matched. Within the brackets we can specify individual characters or, as in this case, ranges of characters; `0-9` means any character between '`0`' and '`9`'.

The plus sign, '`+`', is also a metacharacter, known as a **modifier**. It says that whatever immediately precedes the plus sign in the regular expression can repeat several times. In this case, `[0-9]+` will match one or more digits.

The letters '`r`', '`b`', and the space are all literal, so the entire regular expression will match the letters `rb`, followed by a space, followed by one or more digits. Importantly, this pattern will match *anywhere* within a line; the line does *not* have to begin with `rb`. In other words, this will match rows on which brown rust has been observed on the wheat plant.

Having found which lines contain information about brown rust, we want to extract the information from those lines. The indices from the call to `grep()` can be used to subset out just the relevant lines of data.

```
R> wheat[rbLines]

[1] "lema, rb 2%"   "rb 2%"           "rb 3%"
[4] "rb 4%"         "rb 3%"           "rb 2%,mineuse"
[7] "rb 12"         "rb 25%"          "rj 10, rb 4"
```

We will extract just the brown rust information from these lines in two steps, partly so that we can explore more about regular expressions, and partly because we have to in order to cater for plants that have been afflicted by both brown and yellow rust.

The first step is to reduce the line down to just the information about brown rust. In other words, we want to discard everything *except* the pattern that we are looking for, `rb` followed by a space, followed by one or more digits. The following code performs this step.

```
R> rbOnly <- gsub("^.*(rb [0-9]+).*$", "\\1",
                  wheat[rbLines])
R> rbOnly

[1] "rb 2"  "rb 2"  "rb 3"  "rb 4"  "rb 3"  "rb 2"  "rb 12"
[8] "rb 25" "rb 4"
```

The overall strategy being used here is to match the *entire* line of text, from the first character to the last, but within that line we want to identify and isolate the important component of the line, the brown rust part, so that we can retain just that part.

Again, we have some new metacharacters to explain. First is the "hat" character, '`^`', which matches the start of the line (or the start of the text). Next is the full stop, '`.`'. This will match any single character, no matter what it is. The '`*`' character is similar to the '`+`'; it modifies the immediately preceding part of the expression and allows for *zero* or more occurrences. An expression like `^.*` allows for any number of characters at the start of the text (including zero characters, or an empty piece of text).

The parentheses, '`(`' and '`)`', are used to create **subpatterns** within a regular expression. In this case, we are isolating the pattern `rb [0-9]+`, which matches the brown rust information that we are looking for. Parentheses are useful if we want a modifier, like '`+`' or '`*`', to affect a whole subpattern rather than a single character, *and* they can be useful when specifying the replacement text in a search-and-replace operation, as we will see below.

After the parenthesized subpattern, we have another `.*` expression to allow for any number of additional characters and then, finally, a dollar sign, '$'. The latter is the counterpart to '^'; it matches the *end* of a piece of text.

Thus, the complete regular expression explicitly matches an entire piece of text that contains information on brown rust. Why do we want to do this? Because we are going to replace the entire text with only the piece that we want to keep. That is the purpose of the **backreference**, `"\\1"`, in the replacement text.

The text used to replace a matched pattern in `gsub()` is mostly just literal text. The one exception is that we can refer to subpatterns within the regular expression that was used to find a match. By specifying `"\\1"`, we are saying reuse whatever matched the subpattern within the first set of parentheses in the regular expression. This is known as a **backreference**. In this case, this refers to the brown rust information.

The overall meaning of the `gsub()` call is therefore to replace the entire text with just the part of the text that contains the information about brown rust.

Now that we have character values that contain only the brown rust information, the final step we have to perform is to extract just the numeric data from the brown rust information. We will do this in three ways in order to demonstrate several different techniques.

One approach is to take the text values that contain just the brown rust information and throw away everything except the numbers. The following code does this using a regular expression.

```
R> gsub("[^0-9]", "", rbOnly)

[1] "2"  "2"  "3"  "4"  "3"  "2"  "12" "25" "4"
```

The point about this regular expression is that it uses ^ as the first character within the square brackets. This has the effect of *negating* the set of characters within the brackets, so `[^0-9]` means any character that is *not* a digit. The effect of the complete `gsub()` call is to replace anything that is not a digit with an empty piece of text, so only the digits remain.

An alternative approach is to recognize that the text values that we are dealing with have a very regular structure. In fact, all we need to do is drop the first three characters from each piece of text. The following code does this with a simple call to `substring()`.

```
R> substring(rbOnly, 4)

[1] "2"  "2"  "3"  "4"  "3"  "2"  "12" "25" "4"
```

The first argument to `substring()` is the text to reduce and the second argument specifies which character to start from. In this case, the first character we want is character 4. There is an optional third argument that specifies which character to stop at, but if, as in this example, the third argument is not specified, then we keep going to the end of the text.

The name of this function comes from the fact that text values, or character values, are also referred to as **strings**.

The final approach that we will consider works with the entire original text, `wheat[rbLines]`, and uses a regular expression containing an extra set of parentheses to isolate just the numeric content of the brown rust information as a subpattern of its own. The replacement text refers to this second subpattern, `"\\2"`, so it reduces the entire line to only the part of the line that is the numbers within the brown rust information, in a single step.

```
R> gsub("^.*(rb ([0-9]+)).*$", "\\2", wheat[rbLines])

[1] "2"  "2"  "3"  "4"  "3"  "2"  "12" "25" "4"
```

We are not quite finished because we want to produce a variable that contains the brown rust information for all plants. We will use `NA` for plants that were not afflicted.

A simple way to do this is to create a vector of `NA`s and then fill in the rows for which we have brown rust information. The other important detail in the following code is the conversion of the textual information into numeric values using `as.numeric()`.

```
R> rb <- rep(NA, length(wheat))
R> rb[rbLines] <- as.numeric(gsub("^.*(rb ([0-9]+)).*$",
                                  "\\2", wheat[rbLines]))
R> rb

 [1]  2  2  3  4  3  2 NA NA 12 NA NA NA NA 25 NA NA NA  4
```

To complete the exercise, we need to repeat the process for yellow rust. Rather than repeat the approach used for brown rust, we will investigate a different solution, which will again allow us to demonstrate more text processing techniques.

This time, we will use `regexpr()` rather than `grep()` to find the lines that we want. We are now searching for the lines containing *yellow* rust data.

```
R> rjData <- regexpr("rj [0-9]+", wheat)
R> rjData

 [1] -1 -1 -1 -1 -1 -1 -1 -1 -1 -1  1 -1 -1 -1 -1 -1 -1  1
attr(,"match.length")
 [1] -1 -1 -1 -1 -1 -1 -1 -1 -1 -1  5 -1 -1 -1 -1 -1 -1  5
```

The result is a numeric vector with a positive number for lines that contain yellow rust data and `-1` otherwise. The number indicates the character where the data start. Figure 9.14 includes a diagram that illustrates how this function works.

In this case, there are only two lines containing yellow rust data (lines 11 and 18) and, in both cases, the data start at the first character.

The result also has an attribute called `match.length`, which contains the number of characters that produced the match with the regular expression that we were searching for. In both cases, the pattern matched a total of 5 characters: the letters `r` and `j`, followed by a space, followed by two digits. This length information is particularly useful because it will allow us to extract the yellow rust data immediately using `substring()`. This time we specify both a start and an end character for the subset of the text.

```
R> rjText <- substring(wheat, rjData,
                      attr(rjData, "match.length"))
R> rjText

 [1] ""      ""      ""      ""      ""      ""      ""
 [8] ""      ""      ""      "rj 30" ""      ""      ""
[15] ""      ""      ""      "rj 10"
```

Obtaining the actual numeric data can be carried out using any of the techniques we described above for the brown rust case.

The following code produces the final result, including both brown and yellow rust as a data frame.

```
R> rj <- as.numeric(substring(rjText, 4))
R> data.frame(rb=rb, rj=rj)
```

```
   rb rj
1   2 NA
2   2 NA
3   3 NA
4   4 NA
5   3 NA
6   2 NA
7  NA NA
8  NA NA
9  12 NA
10 NA NA
11 NA 30
12 NA NA
13 NA NA
14 25 NA
15 NA NA
16 NA NA
17 NA NA
18  4 10
```

Recap

Text processing includes: searching within text for a pattern; replacing the text that matches a pattern; splitting text into smaller pieces; combining smaller pieces of text into larger pieces of text; and converting other types of data into text.

9.10 Data display

As we have seen in most of the sections so far in this chapter, most of the tasks that we perform with data, using a programming language, work with the data in RAM. We access data structures that have been stored in RAM and we create new data structures in RAM.

Something that we have largely ignored to this point is how we get to *see* the data on the computer screen.

The purpose of this section is to address the topic of formatting data values for display.

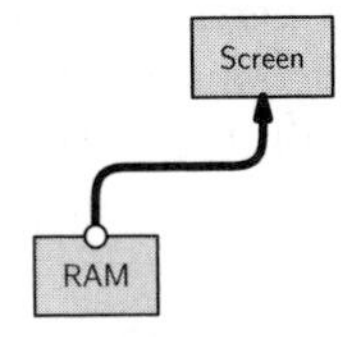

The important thing to keep in mind is that we do not typically see raw computer memory on screen; what we see is a display of the data in a format that is fit for human consumption.

There are two main ways that information is displayed on a computer screen: as *text* output or as *graphical* output (pictures or images). R has sophisticated facilities for graphical output, but that is beyond the scope of this book and we will not discuss those facilities here. Instead, we will focus on displaying text output to the screen.

For a start, we will look a little more closely at how R automatically displays values on the screen.

9.10.1 Case study: Point Nemo (continued)

In this section we will work with a subset of the temperature values from the Point Nemo data set (see Section 1.1).

The temperature values have previously been read into R as the `temp` variable in the data frame called `pointnemodelim` (see Section 9.7.4). In this section, we will only work with the first 12 of these temperature values, which represent the first year's worth of monthly temperature recordings.

```
R> twelveTemps <- pointnemodelim$temp[1:12]
R> twelveTemps
```

```
 [1] 278.9 280.0 278.9 278.9 277.8 276.1 276.1 275.6 275.6
[10] 277.3 276.7 278.9
```

The data structure that we are dealing with is a numeric vector. The values in `twelveTemps` are stored in RAM as numbers.

However, the display that we see on screen is text. The values are numeric, so the characters in the text are mostly digits, but it is important to realize that everything that R displays on screen for us to read is a *text version* of the data values.

The function that displays text versions of data values on screen is called `print()`. This function gets called automatically to display the result of an R expression, but we can also call it directly, as shown below. The display is exactly the same as when we type the name of the symbol by itself.

```
R> print(twelveTemps)
```

```
 [1] 278.9 280.0 278.9 278.9 277.8 276.1 276.1 275.6 275.6
[10] 277.3 276.7 278.9
```

One reason for calling `print()` directly is that this function has arguments that control *how* values are displayed on screen. For example, when displaying numeric values, there is an argument `digits` that controls how many significant digits are displayed.

In the following code, we use the `digits` argument to only display three digits for each temperature value. This has no effect on the values in RAM; it only affects how the numbers are converted to text for display on the screen.

```
R> print(twelveTemps, digits=3)
```

```
 [1] 279 280 279 279 278 276 276 276 276 277 277 279
```

The `print()` function is a generic function (see Section 9.6.8), so what gets displayed on screen is very different for different sorts of data structures; the arguments that provide control over the details of the display will also vary.

Although `print()` has some arguments that control how information is displayed, it is not completely flexible. For example, when printing out a numeric vector, as above, it will always print out the index, in this case [1], at the start of the line.

If we want to have complete control over what gets displayed on the screen, we need to perform the task in two steps: first, generate the text that we want to display, and then call the `cat()` function to display the text on the screen.

For simple cases, the `cat()` function will automatically coerce values to a character vector. For example, the following code uses `cat()` to display the `twelveTemps` numeric vector on screen. The `fill` argument says that a new line should be started after 60 characters have been used.

```
R> cat(twelveTemps, fill=60)

278.9 280 278.9 278.9 277.8 276.1 276.1 275.6 275.6 277.3
276.7 278.9
```

The difference between this display and what `print()` displays is that there is no index at the start of each row. This is the usefulness of `cat()`: it just displays values and does not perform any formatting of its own. This means that we can control the formatting when we generate text values and then just use `cat()` to get the text values displayed on screen.

In summary, the problem of producing a particular display on screen is essentially a problem of generating a character vector in the format that we require and then calling `cat()`.

The next section looks at the problem of generating character vectors in a particular format.

9.10.2 Converting to text

We have previously seen two ways to convert data values to character values: some functions, e.g., `as.character()`, perform an *explicit* type coercion from an original data structure to a character vector; and some functions, e.g., `paste()`, *automatically* coerce their arguments to character vectors. In this section, we will look at some more functions that perform *explicit* coercion to character values.

The following code coerces the `twelveTemps` numeric vector to a character vector using `as.character()`.

```
R> as.character(twelveTemps)

 [1] "278.9" "280"   "278.9" "278.9" "277.8" "276.1" "276.1"
 [8] "275.6" "275.6" "277.3" "276.7" "278.9"
```

One thing to notice about this result is that the second value, "280", is only three characters long, whereas all of the other values are five characters long.

This is a small example of a larger problem that arises when converting values, particularly numbers, to text; there are often many possible ways to perform the conversion. In the case of converting a real number to text, one major problem is how many decimal places to use.

There are several functions in R that we can use to resolve this problem.

The `format()` function produces character values that have a "common format." What that means depends on what sorts of values are being formatted, but in the case of a numeric vector, it means that the resulting character values are all of the same length. In the following result, the second value is five characters long, just like all of the other values.

```
R> format(twelveTemps)

 [1] "278.9" "280.0" "278.9" "278.9" "277.8" "276.1" "276.1"
 [8] "275.6" "275.6" "277.3" "276.7" "278.9"
```

The `format()` function has several arguments that provide some flexibility in the result, but its main benefit is that it displays all values with a common appearance.

For complete control over the conversion to text values, there is the `sprintf()` function.

The following code provides an example of the use of `sprintf()` that converts the `twelveTemps` numeric vector into a character vector where *every* numeric value is converted to a character value with two decimal places and a total of nine characters, followed by a space and a capital letter 'K' (for degrees Kelvin).

```
R> sprintf(fmt="%9.2f K", twelveTemps)

 [1] "   278.90 K" "   280.00 K" "   278.90 K" "   278.90 K"
 [5] "   277.80 K" "   276.10 K" "   276.10 K" "   275.60 K"
 [9] "   275.60 K" "   277.30 K" "   276.70 K" "   278.90 K"
```

The first argument to `sprintf()`, called `fmt`, defines the formatting of the values. The value of this argument can include special codes, like `%9.2f`. The first special code within the `fmt` argument is used to format the second argument to `sprintf()`, in this case the numeric vector `twelveTemps`.

There are a number of special codes for controlling the formatting of different types of values; the components of the format in this example are shown below.

start of special code:	`%9.2f K` (`%`)
real number format:	`%9.2f K` (`f`)
nine characters in total:	`%9.2f K` (`9`)
two decimal places:	`%9.2f K` (`.2`)
literal text:	`%9.2f K` (` K`)

With the `twelveTemps` formatted like this, we can now use `cat()` to display the values on the screen in a format that is quite different from the display produced by `print()`.

```
R> twelveTemps

 [1] 278.9 280.0 278.9 278.9 277.8 276.1 276.1 275.6 275.6
[10] 277.3 276.7 278.9
```

```
R> cat(sprintf("%9.2f K", twelveTemps), fill=60)

   278.90 K    280.00 K    278.90 K    278.90 K    277.80 K
   276.10 K    276.10 K    275.60 K    275.60 K    277.30 K
   276.70 K    278.90 K
```

This sort of formatting can also be useful if we need to generate a plain text file with a particular format. Having generated a character vector as above, this can easily be written to an external text file using the `writeLines()` function or by specifying a filename for `cat()` to write to. If the `file` argument is specified in a call to `cat()`, then output is written to an external file rather than being displayed on the screen.

9.10.3 Results for reports

One reason for using a particular format for displaying results is so that the results can be conveniently included in a research report.

For example, way back in Section 2.1, we saw a very basic web page report about the Pacific and Eurasian Poles of Inaccessibility. The web page is reproduced in Figure 9.16.

This web page includes a summary table showing the range of temperature values for both Poles of Inaccessibility. R code that generates the table of ranges is shown below.

```
R> pointnemotemps <-
       read.fwf(file.path("LAS", "pointnemotemp.txt"),
                skip=8, widths=c(-23, 5),
                col.names=c("temp"))
R> eurasiantemps <-
       read.fwf(file.path("LAS", "eurasiantemp.txt"),
                skip=8, widths=c(-23, 5),
                col.names=c("temp"))
R> allTemps <- cbind(pacific=pointnemotemps$temp,
                     eurasian=eurasiantemps$temp)
R> ranges <- round(rbind(min=apply(allTemps, 2, min),
                         max=apply(allTemps, 2, max)))
R> ranges
```

```
    pacific eurasian
min     276      252
max     283      307
```

The HTML code that includes this table in the web page is reproduced below. This is the most basic way to make use of R output; simply copy the R display directly into another document, with a monospace font.

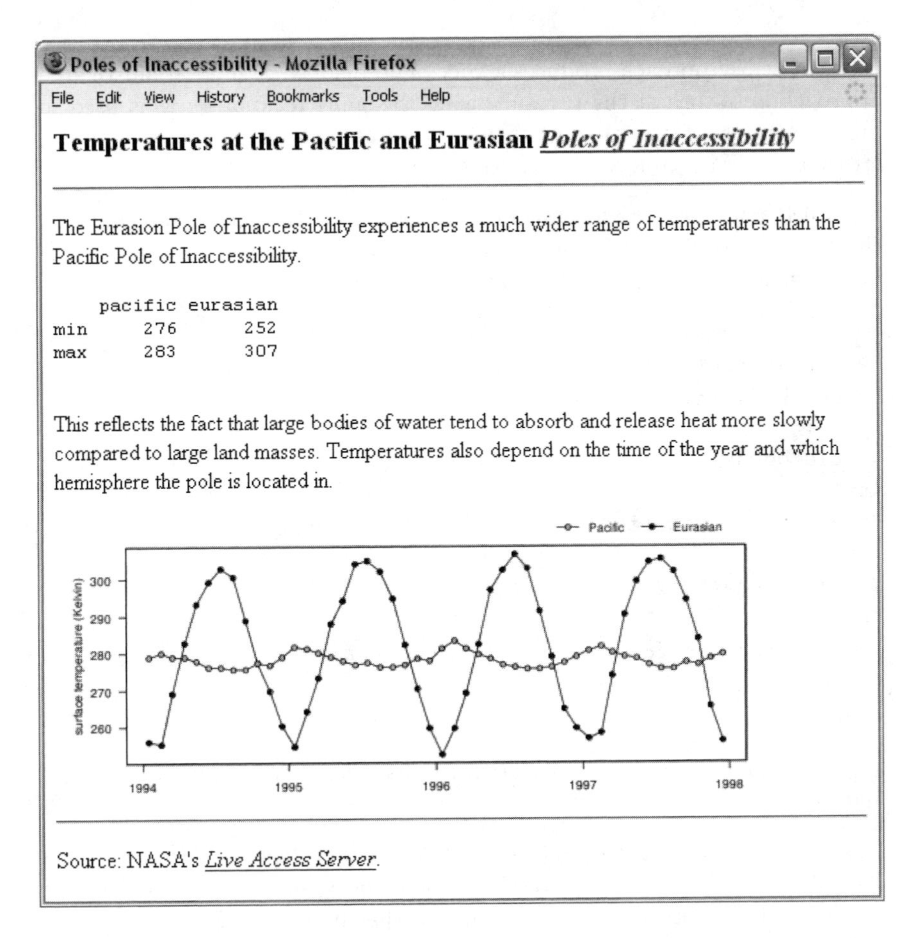

Poles of Inaccessibility - Mozilla Firefox

File Edit View History Bookmarks Tools Help

Temperatures at the Pacific and Eurasian *Poles of Inaccessibility*

The Eurasion Pole of Inaccessibility experiences a much wider range of temperatures than the Pacific Pole of Inaccessibility.

	pacific	eurasian
min	276	252
max	283	307

This reflects the fact that large bodies of water tend to absorb and release heat more slowly compared to large land masses. Temperatures also depend on the time of the year and which hemisphere the pole is located in.

Source: NASA's *Live Access Server*.

Figure 9.16: A simple web page that displays information about the surface temperature data for the Pacific and Eurasian Poles of Inaccessibility (viewed with the Firefox web browser on Windows XP). This is a reproduction of Figure 2.1.

```
     <pre>
     pacific eurasian
min      276      252
max      283      307
     </pre>
```

However, this approach produces a very plain display. A more sophisticated approach is to format the R result using the same technology as is used to produce the report. For example, in the case of a web page report, we could create an HTML table to display the result.

Several R packages provide functions to carry out this task. For example, the **hwriter** package has an `hwrite()` function that converts an R table into text describing an HTML table.

```
R> library(hwriter)
R> cat(hwrite(ranges))

<table border="1">
<tr>
<td></td><td>pacific</td><td>eurasian</td></tr>
<tr>
<td>min</td><td>276</td><td>252</td></tr>
<tr>
<td>max</td><td>283</td><td>307</td></tr>
</table>
```

This approach allows us to integrate R results within a report more naturally and more aesthetically.

It is worth noting that this is just a text processing task; we are converting the values from the R table into text values and then combining those text values with HTML tags, which are just further text values.

This is another important advantage of carrying out tasks by writing computer code; we can use a programming language to write computer code. We can write code to generate our instructions to the computer, which is a tremendous advantage if the instructions are repetitive, for example, if we write the same HTML report every month.

Another option for producing HTML code from R data structures is the **xtable** package; this package can also format R data structures as LaTeX tables.

Although a detailed description is beyond the scope of this book, it is also worth mentioning the **Sweave** package, which allows HTML (or LaTeX) code to be combined with R code within a single document, thereby avoiding having to cut-and-paste results from R to a separate report document by hand.

Recap

When we view data values that have been stored in RAM, what we see displayed on screen is typically a text version of the data values.

The conversion of data values to a text representation is sometimes ambiguous and requires us to provide a specification of what the result should be.

It is possible to format the display of R *data structures so that they can be integrated nicely within research reports.*

9.11 Programming

A computer program is often described as "data structures + algorithms."

Most of this chapter has been concerned with data structures—how information is stored in computer memory—and a number of tools that allow us to convert one data structure into another sort of data structure in a single step.

Algorithms are the series of steps that we take in order to carry out more complex tasks. Algorithms are what we use to combine smaller tasks into computer programs.

The difference between data structures and algorithms is like the difference between the *ingredients* that we need to cook a meal and the *recipe* that we use to cook the meal. The focus of this book is on data structures. Much of what we need to achieve when processing a data set can be performed in only a few steps, with only one or two expressions. This book is mostly about boiling an egg rather than baking a soufflé.

However, as we perform more complex data processing tasks, it becomes useful to know something about how to combine and manage a larger number of expressions.

In this section, we will look at some slightly more advanced ideas for writing

```
             VARIABLE : Mean Near-surface air temperature (kelvin)
             FILENAME : ISCCPMonthly_avg.nc
             FILEPATH : /usr/local/fer_data/data/
             SUBSET   : 24 by 24 points (LONGITUDE-LATITUDE)
             TIME     : 16-JAN-1995 00:00
              113.8W 111.2W 108.8W 106.2W 103.8W 101.2W 98.8W  ...
               27     28     29     30     31     32     33    ...
 36.2N / 51:  272.1  270.3  270.3  270.9  271.5  275.6  278.4  ...
 33.8N / 50:  282.2  282.2  272.7  272.7  271.5  280.0  281.6  ...
 31.2N / 49:  285.2  285.2  276.1  275.0  278.9  281.6  283.7  ...
 28.8N / 48:  290.7  286.8  286.8  276.7  277.3  283.2  287.3  ...
 26.2N / 47:  292.7  293.6  284.2  284.2  279.5  281.1  289.3  ...
 23.8N / 46:  293.6  295.0  295.5  282.7  282.7  281.6  285.2  ...
 ...
```

Figure 9.17: The first few lines of output from the Live Access Server for the near-surface air temperature of the earth for January 1995, over a coarse 24 by 24 grid of locations covering Central America.

computer code. This treatment will barely scratch the surface of the topics available; the aim is to provide a very brief introduction to some useful ideas for writing larger amounts of more complex programming code.

We will use a case study to motivate the need for these more advanced techniques.

9.11.1 Case study: The Data Expo (continued)

The data for the 2006 JSM Data Expo (Section 5.2.8) were obtained from NASA's Live Access Server as a set of 505 text files.

Seventy-two of those files contain near-surface air temperature measurements, with one file for each month of recordings. Each file contains average temperatures for the relevant month at 576 different locations. Figure 9.17 shows the first few lines of the temperature file for the first month, January 1995.

With the data expo files stored in a local directory, `NASA/Files`, the complete set of 72 filenames for the files containing temperature recordings can be generated by the following code (only the first six filenames are shown).

```
R> nasaAirTempFiles <-
       file.path("NASA", "Files",
                 paste("temperature", 1:72, ".txt",
                       sep=""))
R> head(nasaAirTempFiles)

[1] "NASA/Files/temperature1.txt"
[2] "NASA/Files/temperature2.txt"
[3] "NASA/Files/temperature3.txt"
[4] "NASA/Files/temperature4.txt"
[5] "NASA/Files/temperature5.txt"
[6] "NASA/Files/temperature6.txt"
```

We will conduct a simple task with these data: calculating the near-surface air temperature for each month, averaged over all locations. In other words, we will calculate a single average temperature from each file. The result will be a vector of 72 monthly averages.

As a starting point, the following code reads in the data from just a single file, using the first filename, `temperature1.txt`, and averages all of the temperatures in that file.

```
R> tempDF <- read.fwf(nasaAirTempFiles[1],
                      skip=7,
                      widths=c(-12, rep(7, 24)))
R> tempMatrix <- as.matrix(tempDF)
R> mean(tempMatrix)

[1] 295.1849
```

The call to `read.fwf()` ignores the first 7 lines of the file and the first 12 characters on each of the remaining lines of the file. This just leaves the temperature values, which are stored in RAM as a data frame, `tempDF`. This data frame is then converted into a matrix, `tempMatrix`, so that, in the final step, the `mean()` function can calculate the average across all of the temperature values.

At this point, we have simple code that uses the functions and data structures that have been described in previous sections. The next step is to repeat this task for all 72 air temperature files.

One way to do that is to write out 72 almost identical copies of the same piece of code, but, unsurprisingly, there are ways to work much more efficiently. The next section describes how.

9.11.2 Control flow

In a programming language, if our code contains more than one expression, the expressions are run one at a time, in the order that they appear.

For example, consider the following two expressions from the code in the previous section.

```
tempMatrix <- as.matrix(tempDF)
mean(tempMatrix)
```

In this code, the expression `mean(tempMatrix)` relies on the fact that the previous expression has already been run and that there is a matrix data structure stored in RAM for it to access via the symbol `tempMatrix`.

However, programming languages also provide ways to modify this basic rule and take control of the order in which expressions are run. One example of this is the idea of a `loop`, which is a way to allow a collection of expressions to be run repeatedly.

Returning to our example of calculating an average value for each of 72 different files, the following code shows how to calculate these averages using a loop in R.

```
R> avgTemp <- numeric(72)
R> for (i in 1:72) {
       tempDF <- read.fwf(nasaAirTempFiles[i],
                          skip=7,
                          widths=c(-12,
                            rep(7, 24)))
       tempMatrix <- as.matrix(tempDF)
       avgTemp[i] <- mean(tempMatrix)
   }
```

The first expression is just a set-up step that creates a vector of 72 zeroes. This vector will be used to store the 72 average values as we calculate them.

The components of the actual loop are shown below.

```
        keywords:  for (i in 1:72) {
     parentheses:  for (i in 1:72) {
     loop symbol:  for (i in 1:72) {
loop symbol values: for (i in 1:72) {
    open bracket:  for (i in 1:72) {
       loop body:      tempDF <- read.fwf(nasaAirTempFiles[i],
                                          skip=7,
                                          widths=c(-12,
                                            rep(7, 24)))
                       tempMatrix <- as.matrix(tempDF)
                       avgTemp[i] <- mean(tempMatrix)
   close bracket:  }
```

The keywords, `for` and `in`, the brackets, and the parentheses will be the same in any loop that we write. The most important bits are the loop symbol, in this case the symbol `i`, and the values that the loop symbol can take, in this case, `1:72`.

The idea of a loop is that the expressions in the body of the loop are run several times. Specifically, the expressions in the body of the loop are run once for each value of the loop symbol. Also, each time the loop repeats, a different value is assigned to the loop symbol.

In this example, the value `1` is assigned to the symbol `i` and the expressions in the body of the loop are run. Then, the value `2` is assigned to the symbol `i` and the body of the loop is run again. This continues until we assign the value `72` to the symbol `i` and run the body of the loop, and then the loop ends.

If we look closely at the expressions in the body of the loop, we can see that the loop symbol, `i`, is used twice. In the first case, `nasaAirTempFiles[i]`, this means that each time through the loop we will read a different air temperature file. In the second case, `avgTemp[i]`, this means that the average that we calculate from the air temperature file will be stored in a different element of the vector `avgTemp`, so that by the end of the loop, we have all 72 averages stored together in the vector.

The important idea is that, although the code in the body of the loop remains constant, the code in the body of the loop can do something slightly different each time the loop runs because the loop symbol changes its value every time the loop runs.

The overall result is that we read all 72 files, calculate an average from each one, and store the 72 averages in the vector avgTemp, which is shown below.

```
R> avgTemp

 [1] 295.1849 295.3175 296.3335 296.9587 297.7286 298.5809
 [7] 299.1863 299.0660 298.4634 298.0564 296.7019 295.9568
[13] 295.3915 296.1486 296.1087 297.1007 298.3694 298.1970
[19] 298.4031 298.0682 298.3148 297.3823 296.1304 295.5917
[25] 295.5562 295.6438 296.8922 297.0823 298.4793 299.3575
[31] 299.7984 299.7314 299.6090 298.4970 297.9872 296.8453
[37] 296.9569 296.9354 297.0240 298.0668 299.1821 300.7290
[43] 300.6998 300.3715 300.1036 299.2269 297.8642 297.2729
[49] 296.8823 297.4288 297.5762 298.2859 299.1076 299.1938
[55] 299.0599 299.5424 298.9135 298.2849 297.0981 296.2639
[61] 296.1943 296.5868 297.5510 298.6106 299.7425 299.5219
[67] 299.7422 300.3411 299.5781 298.6965 297.0830 296.3813
```

9.11.3 Writing functions

When we have to write code to perform the same task several times, one approach, as described in the previous section, is to write a loop.

In this section, we will look at another option: writing functions.

In our example, we want to calculate an overall average for each of 72 separate files. One way to look at this task is that there are 72 filenames stored in a character vector data structure and we want to perform a calculation for each element of this data structure. As we saw in Section 9.8.7, there are several R functions that are designed to perform a task on each element of a data structure.

What we can do in this case is use sapply() to call a function once for each of the filenames in the vector nasaAirTempFiles.

The call to the sapply() function that we would like to make is shown below.

```
sapply(nasaAirTempFiles, calcMonthAvg)
```

The bad news is that the function calcMonthAvg() does not exist! There is no predefined function that can read a NASA air temperature file and return the average of all of the values in the file.

The good news is that we can very easily create this function ourselves.

Here is code that creates a function called `calcMonthAvg()` to perform this task.

```
R> calcMonthAvg <- function(filename) {
       tempDF <- read.fwf(filename,
                          skip=7,
                          widths=c(-12, rep(7, 24)))
       tempMatrix <- as.matrix(tempDF)
       mean(tempMatrix)
   }
```

This code creates a **function** and assigns it to the symbol `calcMonthAvg`. In most examples in this chapter, we have stored *data structures* in RAM, but in this case, we have stored a *function* in RAM; we have stored some computer code in RAM. Just like with any other symbol, we can retrieve the function from RAM by referring to the symbol `calcMonthAvg`.

```
R> calcMonthAvg
```

```
function (filename)
{
    tempDF <- read.fwf(filename, skip = 7,
        widths = c(-12, rep(7, 24)))
    tempMatrix <- as.matrix(tempDF)
    mean(tempMatrix)
}
```

However, because this is a function, we can do more than just retrieve its value; we can also *call* the function. This is done just like any other function call we have ever made. We append parentheses to the symbol name and include arguments within the parentheses. The following code calls our new function to calculate the average for the first air temperature file.

```
R> calcMonthAvg(nasaAirTempFiles[1])
```

```
[1] 295.1849
```

In order to understand what happens when we call our function, we need to take a closer look at the components of the code that was used to create the function.

```
    function name:  calcMonthAvg <- function(filename) {
          keyword:  calcMonthAvg <- function(filename) {
      parentheses:  calcMonthAvg <- function(i) {
  argument symbol:  calcMonthAvg <- function(i) {
     open bracket:  calcMonthAvg <- function(filename) {
    function body:      tempDF <- read.fwf(filename,
                                           skip=7,
                                           widths=c(-12,
                                                    rep(7, 24)))
                        tempMatrix <- as.matrix(tempDF)
                        mean(tempMatrix)
    close bracket:  }
```

The keyword `function`, the brackets, and the parentheses will be the same for every function that we write. The important bits are the name we choose for the function and the argument symbols that we choose for the function.

The idea of a function is that the expressions that make up the body of the function will be run when the function is called. Also, when the function is called, the value that is provided as the argument in the function call will be assigned to the argument symbol.

In this case, the argument symbol is `filename` and this symbol is used once in the body of the function, in the call to `read.fwf()`. This means that the argument to the function is used to select which file the function will read and calculate an average value for.

The value returned by a function is the value of the last expression within the function body. In this case, the return value is the result of the call to the `mean()` function.

As a simple example, the following code is a call to our `calcMonthAvg()` function that calculates the overall average temperature from the contents of the file `temperature1.txt`.

```
R> calcMonthAvg(nasaAirTempFiles[1])
```

```
[1] 295.1849
```

That function call produces exactly the same result as the following code.

```
R> tempDF <- read.fwf(nasaAirTempFiles[1],
                      skip=7,
                      widths=c(-12, rep(7, 24)))
R> tempMatrix <- as.matrix(tempDF)
R> mean(tempMatrix)
```

The advantage of having defined the function `calcMonthAvg()` is that it is now possible to calculate the monthly averages from all of the temperature files using `sapply()`.

We supply the vector of filenames, `nasaAirTempFiles`, and our new function, `calcMonthAvg()`, and `sapply()` calls our function for each filename. The `USE.NAMES` argument is employed here to avoid having large, messy names on each element of the result.

```
R> sapply(nasaAirTempFiles, calcMonthAvg,
          USE.NAMES=FALSE)

 [1] 295.1849 295.3175 296.3335 296.9587 297.7286 298.5809
 [7] 299.1863 299.0660 298.4634 298.0564 296.7019 295.9568
[13] 295.3915 296.1486 296.1087 297.1007 298.3694 298.1970
[19] 298.4031 298.0682 298.3148 297.3823 296.1304 295.5917
[25] 295.5562 295.6438 296.8922 297.0823 298.4793 299.3575
[31] 299.7984 299.7314 299.6090 298.4970 297.9872 296.8453
[37] 296.9569 296.9354 297.0240 298.0668 299.1821 300.7290
[43] 300.6998 300.3715 300.1036 299.2269 297.8642 297.2729
[49] 296.8823 297.4288 297.5762 298.2859 299.1076 299.1938
[55] 299.0599 299.5424 298.9135 298.2849 297.0981 296.2639
[61] 296.1943 296.5868 297.5510 298.6106 299.7425 299.5219
[67] 299.7422 300.3411 299.5781 298.6965 297.0830 296.3813
```

This is the same as the result that was produced from an explicit loop (see page 354), but it uses only a single call to `sapply()`.

9.11.4 Flashback: Writing functions, writing code, and the DRY principle

The previous example demonstrates that it is useful to be able to define our own functions for use with functions like `apply()`, `lapply()`, and `sapply()`. However, there are many other good reasons for being able to write functions. In particular, functions are useful for organizing code, simplifying code, and for making it easier to maintain code.

For example, the code below reproduces the loop that we wrote in Section 9.11.2, which calculates all 72 monthly averages.

```
for (i in 1:72) {
    tempDF <- read.fwf(nasaAirTempFiles[i],
                       skip=7,
                       widths=c(-12,
                         rep(7, 24)))
    tempMatrix <- as.matrix(tempDF)
    avgTemp[i] <- mean(tempMatrix)
}
```

We can use the `calcMonthAvg()` function to make this code much simpler and easier to read.

```
for (i in 1:72) {
    avgTemp[i] <- calcMonthAvg(nasaAirTempFiles[i])
}
```

This sort of use for functions is just an extension of the ideas of laying out and documenting code for the benefit of human readers that we saw in Section 2.4.

A further advantage that we can obtain from writing functions is the ability to *reuse* our code.

The `calcMonthAvg()` function nicely encapsulates the code for calculating the overall average temperature from one of these NASA files. If we ever need to perform this calculation in another analysis, rather than writing the code again, we can simply make use of this function again.

In this sense, functions are another example of the DRY principle because a function allows us to create and maintain a single copy of a piece of information—in this case, the computer code to perform a specific task.

9.11.5 Flashback: Debugging

All of the admonitions about writing code in small pieces and changing one thing at a time, which were discussed first in Section 2.6.2, are even more important when writing code in a programming language because the code tends to be more complex.

A specific point that can be made about writing R code, especially code that involves many expressions, is that each individual expression can be run, one at a time, in the R environment. If the final result of a collection of R code is not correct, the first thing to do is to run the individual expressions, one at a time, in order, and check that the result of each expression is correct. This is an excellent way to track down where a problem is occurring.

Matters become far more complicated once we begin to write our own functions, or even if we write loops. If the code within a function is not performing as expected, the `debug()` function is very useful because it allows us to run the code *within a function* one expression at a time.

Another useful simple trick, for both functions and loops, is to use the functions from Section 9.10 to write messages to the screen that show the values from intermediate calculations.

Recap

A loop is an expression that repeats a body of code multiple times.

Functions allow a piece of code to be reused.

In their simplest form, functions are like macros, simply a recording of a series of steps.

The best way to debug R *code is to run it one expression at a time.*

9.12 Other software

This chapter has focused on the R language for simple programming tasks.

The tools described in this chapter are the core tools for working with the fundamental data structures in R. In specific areas of research, particularly where data sets have a special format, there may be R packages that provide more sophisticated and convenient tools for working with a specific data format.

A good example of this is the **zoo** package for working with time series data. Other examples are the packages within the Bioconductor project[4] that provide tools for working with the results of microarray experiments.

The choice of R as a data processing tool was based on the fact that R is

[4]`http://www.bioconductor.org/`

a relatively easy programming language to learn and it has good facilities for manipulating data. R is also an excellent environment for data *analysis*, so learning to process data with R means that it is possible to prepare *and* analyze a data set within a single system.

However, there are two major disadvantages to working with data using R: R is slower than many other programming languages, and R holds all data in RAM, so it can be awkward to work with extremely large data sets.

There are a number of R packages that enhance R's ability to cope gracefully with very large data sets. One approach is to store the data in a relational database and use the packages that allow R to talk to the database software, as described in Section 9.7.8. Several other packages solve the problem by storing data in mass storage rather than RAM and just load data as required into RAM; two examples are the **filehash** and **ff** packages.

If R is too slow for a particular task, it may be more appropriate to use a different data processing tool.

There are many alternative programming languages, such as C, Perl, Python, and Java, that will run much faster than R. The trade-off with using one of these programming languages is likely to involve writing more code, and more complex code, in order to produce code that runs faster.

It is also worth mentioning that many simple software tools exist, especially for processing text files. In particular, on Linux systems, there are programs such as `sort` for sorting files, `grep` for finding patterns in files, `cut` for removing pieces from each line of a file, and `sed` and `awk` for more powerful rearrangements of file contents. These tools are not as flexible or powerful as R, but they are extremely fast and will work on files of virtually any size. The Cygwin project[5] makes these tools available on Windows.

[5] `http://www.cygwin.com/`

Summary

A programming language provides great flexibility and power because it can be used to control the computer hardware, including accessing data from mass storage, processing data values within RAM, and displaying data on screen.

R *is a good programming language for working with data because it is relatively easy to learn and it provides many tools for processing data.*

A data structure is the format in which a collection of data values is stored together within RAM. The basic data structures in R *are vectors, matrices, data frames, and lists.*

Standard data processing tasks include: reading data from external files into RAM; extracting subsets of data; sorting, transforming, aggregating, and reshaping data structures; searching and reformatting text data; and formatting data for display on screen.

10

R Reference

R is a popular programming language and interactive environment for processing and analyzing data.

This chapter provides brief reference information for the R *language*. See Section 9.2 for a brief introduction to the R *software* environment that is used to run R code.

10.1 R syntax

R code consists of one or more **expressions**. This section describes several different sorts of expressions.

10.1.1 Constants

The simplest type of expression is just a constant value. The most common constant values in R are numbers and text. There are various ways to enter numbers, including using scientific notation and hexadecimal syntax. Text must be surrounded by double-quotes or single-quotes, and special characters may be included within text using various escape sequences. The help pages `?NumericConstants` and `?Quotes` provide a detailed description of the various possibilities.

Any constant not starting with a number and not within quotes is a **symbol**.

There are a number of reserved symbols with predefined meanings: `NA` (missing value), `NULL` (an empty data structure), `NaN` (Not a Number), `Inf` and `-Inf` ([minus] infinity), `TRUE` and `FALSE`, and the symbols used for control flow as described below.

Section 10.1.5 will describe how to create new symbols.

10.1.2 Arithmetic operators

R has all of the standard arithmetic operators such as addition (+), subtraction (-), division (/), multiplication (*), and exponentiation (^). R also has operators for integer division (%/%) and remainder on integer division (%%; also known as modulo arithmetic).

10.1.3 Logical operators

The comparison operators <, >, <=, >=, ==, and != are used to determine whether values in one vector are larger or smaller or equal to the values in another vector. The %in% operator determines whether each value in the left operand can be matched with one of the values in the right operand. The result of these operators is a logical vector.

The logical operators || (or) and && (and) can be used to combine two logical values and produce another logical value as the result. The operator ! (not) negates a logical value. These operators allow complex conditions to be constructed.

The operators | and & are similar, but they combine two logical *vectors*. The comparison is performed element by element, so the result is also a logical vector.

Section 10.3.4 describes several functions that perform comparisons.

10.1.4 Function calls

A function call is an expression of the form:

```
functionName(arg1, arg2)
```

A function can have any number of arguments, including zero. Every argument has a name.

Arguments can be specified by position or by name (name overrides position). Arguments may have a default value, which they will take if no value is supplied for the argument in the function call.

All of the following function calls are equivalent (they all generate a numeric vector containing the integers 1 to 10):

```
seq(1, 10)               # positional arguments
seq(from=1, to=10)       # named arguments
seq(to=10, from=1)       # names trump position
seq(1, 10, by=1)         # 'by' argument has default
```

Section 10.3 provides details about a number of important functions for basic data processing.

10.1.5 Symbols and assignment

Anything not starting with a digit, that is not a special keyword, is treated as a symbol. Values may be assigned to symbols using the `<-` operator; otherwise, any expression involving a symbol will produce the value that has been previously assigned to that symbol.

```
R> x <- 1:10
```

```
R> x
```

```
 [1]  1  2  3  4  5  6  7  8  9 10
```

10.1.6 Loops

A loop is used to repeatedly run a group of expressions.

A `for` loop runs expressions a fixed number of times. It has the following general form:

```
for (symbol in sequence) {
    expressions
}
```

The *expressions* are run once for each element in the *sequence*, with the relevant element of the *sequence* assigned to the *symbol*.

A `while` loop runs expressions until a condition is met. It has the following general form:

```
while (condition) {
    expressions
}
```

The `while` loop repeats until the `condition` is `FALSE`. The `condition` is an expression that should produce a single logical value.

10.1.7 Conditional expressions

A conditional expression is used to make expressions contingent on a condition.

A conditional expression in R has the following form:

```
if (condition) {
    expressions
}
```

The *condition* is an expression that should produce a single logical value, and the *expressions* are only run if the result of the *condition* is `TRUE`.

The curly braces are not necessary, but it is good practice to always include them; if the braces are omitted, only the first complete expression following the `condition` is run.

It is also possible to have an `else` clause.

```
if (condition) {
    trueExpressions
} else {
    falseExpressions
}
```

10.2 Data types and data structures

Individual values are either **character** values (text), **numeric** values (numbers), or **logical** values (`TRUE` or `FALSE`). R also supports complex values with an imaginary component.

There is a distinction within numeric values between integers and real values, but integer values tend to be coerced to real values if anything is done to them. If an integer is required, it is best to use a function that explicitly generates integer values, such as `as.integer()`.

On a 32-bit operating system, in an English locale, a character value uses 1 byte per character; an integer uses 4 bytes, as does a logical value; and a real number uses 8 bytes. The function `object.size()` returns the approximate number of bytes used by an R data structure in memory.

```
R> object.size(1:1000)

[1] 4024
```

```
R> object.size(as.numeric(1:1000))

[1] 8024
```

The simplest data structure in R is a vector. All elements of a vector must have the same basic type. Most operators and many functions accept vector arguments and return a vector result.

Matrices and arrays are multidimensional analogues of the vector. All elements must have the same type.

Data frames are collections of vectors where each vector must have the same length, but different vectors can have different types. This data structure is the standard way to represent a data set in R.

Lists are like vectors that can have different types of data structures in each component. In the simplest case, each component of a list may be a vector of values. Like the data frame, each component can be a vector of a different basic type, but for lists there is no requirement that each component has the same size. More generally, the components of a list can be more complex data structures, such as matrices, data frames, or even other lists. Lists can be used to efficiently represent hierarchical data in R.

10.3 Functions

This section provides a list of some of the functions that are useful for working with data in R. The descriptions of these functions are very brief and only some of the arguments to each function are mentioned. For a complete description of the function and its arguments, the relevant function help page should be consulted (see Section 10.4).

10.3.1 Session management

This section describes some functions that are useful for querying and controlling the R software environment during an interactive session.

`ls()`

List the symbols that have had values assigned to them during the current session.

`rm(...)`
`rm(list)`

Delete one or more symbols (the value that was assigned to the symbol is no longer accessible). The symbols to delete are specified by name *or* as a list of names.

To delete all symbols in the current session, use `rm(list=ls())` (carefully).

`options(...)`

Set a global option for the R session by specifying a new value with an appropriate argument name in the form *optionName=optionValue* *or* query the current setting for an option by specifying "*optionName*".

Typing `options()` with no arguments returns a list of all current option settings.

`q()`

Exit the current R session.

10.3.2 Generating vectors

`c(...)`

Create a new vector by concatenating or combining the values (or vectors of values) given as arguments. All values must be of the same type (or they will be coerced to the same type).

This function can also be used to concatenate lists.

`seq(from, to, by)`
`seq(from, to, length.out)`
`seq(from, to, along.with)`

Generate a sequence of numbers from the value `from` to (not greater than) the value `to` in steps of `by`, *or* for a total of `length.out` values, *or* so that the sequence has the same length as `along.with`.

The function `seq_len(n)` is faster for producing the sequence from 1 to `n` and `seq_along(x)` is faster for producing the sequence from 1 to the number of values in `x`. These may be useful for producing very long sequences.

The colon operator, `:`, provides a short-hand syntax for sequences of integer values in steps of 1. The expression *from*`:`*to* is equivalent to `seq(`*from*`, `*to*`)`.

`rep(x, times)`
`rep(x, each)`
`rep(x, length.out)`

Repeat all values in a vector `times` times, *or* each value in the vector `each` times, *or* all values in the vector until the total number of values is `length.out`.

`append(x, values, after)`

Insert the `values` into the vector `x` at the position specified by `after`.

`unlist(x)`

Convert a list structure into a vector by concatenating all components of the list. This is especially useful when a function call returns a list where each component is a vector.

`rev(x)`

Reverse the elements of a vector.

`unique(x)`

Remove any duplicated values from `x`.

R

10.3.3 Numeric functions

`sum(..., na.rm=FALSE)`

Sum the value of all arguments. If `NA` values are included, the result is `NA` (unless `na.rm=TRUE`).

`mean(x)`

Calculate the arithmetic mean of the values in `x`.

`max(..., na.rm=FALSE)`
`min(..., na.rm=FALSE)`
`range(..., na.rm=FALSE)`

Calculate the minimum, maximum, or range of all values in all arguments.

The functions `which.min()` and `which.max()` return the index of the minimum or maximum value within a vector.

`diff(x)`
: Calculate the difference between successive values of `x`. The result contains one fewer values than there are in `x`.

`cumsum(x)`
`cumprod(x)`
: The cumulative sum or cumulative product of the values in `x`.

10.3.4 Comparisons

`identical(x, y)`
: Tests whether `x` and `y` are equivalent down to the details of their representation in computer memory.

`all.equal(target, current, tolerance)`
: Tests whether `target` and `current` differ by only a tiny amount, where "tiny" is defined by `tolerance`). This is useful for testing whether numeric values are equal.

`match(x, table)`
: Determine the location of each element of `x` in the set of values in `table`. The result is a numeric index the same length as `x`.

The `%in%` operator is similar (`x %in% table`), but returns a logical vector the same length as `x` reporting whether each element of `x` was found in `table`.

The `pmatch()` function performs partial matching (whereas `match()` is exact).

`is.null(x)`
`is.na(x)`
`is.infinite(x)`
`is.nan(x)`
: These functions should be used to test for the special values `NULL`, `NA`, `Inf`, and `NaN`.

`any(...)`
`all(...)`
: Test whether all or any values in one or more logical vectors are `TRUE`. The result is a single logical value.

10.3.5 Type coercion

```
as.character(x)
as.logical(x)
as.numeric(x)
as.integer(x)
```

Convert the data structure `x` to a vector of the appropriate type.

```
as.Date(x, format)
as.Date(x, origin)
```

Convert character values or numeric values to Date values.

Character values are converted automatically if they are in ISO 8601 format; otherwise, it may be necessary to describe the date format via the `format` argument. The help page for the `strftime()` function describes the syntax for specifying date formats.

When converting numeric values, a reference date must be provided, via the `origin` argument.

The `Sys.Date()` function returns today's date as a date value.

The `months()` function resolves date values just to month names. There are also functions for `weekdays()` and `quarters()`.

```
floor(x)
ceiling(x)
round(x, digits)
```

Round a numeric vector, `x`, to `digits` decimal places or to an integer value. `floor()` returns largest integer not greater than `x` and `ceiling()` returns smallest integer not less than `x`.

```
signif(x, digits)
```

Round a numeric vector, `x`, to `digits` significant digits.

10.3.6 Exploring data structures

```
attributes(x)
attr(x, which)
```

Extract a list of all attributes, or just the attributes named in the character vector `which`, from the data structure `x`.

`names(x)`
`rownames(x)`
`colnames(x)`
`dimnames(x)`

Extract the names attribute from a vector or list, or the row names or column names from a two-dimensional data structure, or the list of names for all dimensions of an array.

`summary(object)`
Produces a summary of `object`. The information displayed will depend on the class of `object`.

`length(x)`
The number of elements in a vector, or the number of components in a list. Also works for data frames and matrices, though the result may be less intuitive; it gives the number of columns for a data frame and the *total* number of values in a matrix.

`dim(x)`
`nrow(x)`
`ncol(x)`

The dimensions of a matrix, array, or data frame. `nrow()` and `ncol()` are specifically for two-dimensional data structures, but `dim()` will also work for higher-dimensional structures.

`head(x, n=6)`
`tail(x, n=6)`

Return just the first or last `n` elements of a data structure; the first elements of a vector, the first few rows of a data frame, and so on.

`class(x)`
Return the class of the data structure `x`.

`str(object)`
Display a summarized, low-level view of a data structure. Typically, the output is less pretty and more detailed than the output from `summary()`.

10.3.7 Subsetting

Subsetting is generally performed via the square bracket operator, `[` (e.g., `candyCounts[1:4]`). In general, the result is of the same class as the original data structure that is being subsetted. The subset may be a numeric vector, a character vector (names), or a logical vector (the same length as the original data structure).

When subsetting data structures with more than one dimension—e.g., data frames, matrices, or arrays—the subset may be several vectors, separated by commas (e.g., `candy[1:4, 4]`).

The double square bracket operator, `[[`, selects only one component of a data structure. This is typically used to extract a component from a list.

`subset(x, subset, select)`
: Extract the rows of the data frame `x` that satisfy the condition in `subset` and the columns that are named in `select`.

An important special case of subsetting for statistical data sets is the issue of removing missing values from a data set. The function `na.omit()` can be used to remove all rows containing missing values from a data frame.

10.3.8 Data import/export

R provides general functions for working with the file system.

`getwd()`
`setwd(dir)`
: Get the current working directory or set it to `dir`. This is where R will look for files (or start looking for files).

`list.files(path, pattern)`
: List the names of files in the directory given by `path`, filtering results with the specified `pattern` (a regular expression).

  For Linux users who are used to using filename globs with the `ls` shell command, this use of regular expressions for filename patterns can cause confusion. Such users may find the `glob2rx()` function helpful.

  The complete names of the files, including the path, can be obtained by specifying `full.names=TRUE`. Given a full filename, consisting of a path and a filename, `basename()` strips off the path to leave just the filename, and `dirname()` strips off the filename to leave just the path.

`file.path(...)`
: Given the names of nested directories, combine them using an appropriate separator to form a path.

`file.choose()`
: Interactively select a file (on Windows, using a dialog box interface).

`file.exists()`
`file.remove()`
`file.rename()`
`file.copy()`
`dir.create()`

These functions perform the standard file manager tasks of copying, deleting, and renaming files and creating new directories.

There are a number of functions for reading data from external text files into R.

`readLines(con)`

Read the text file specified by the filename or path given by `con`. The file specification can also be a URL. The result is a character vector with one element for each line in the file.

`read.table(file, header=FALSE, skip=0, sep="")`

Read the text file specified by the character value in `file`, treating each line of text as a case in a data set that contains values for each variable in the data set, with values separated by the character value in `sep`. Ignore the first `skip` lines in the file. If `header` is `TRUE`, treat the first line of the file as variable names.

The default behavior is to treat columns that contain *only* numbers as numeric and to treat everything else as a factor. The arguments `as.is` and `stringsAsFactors` can be used to produce character variables rather than factors. The `colClasses` argument provides further control over the type of each column.

This function can be slow on large files because of the work it does to determine the type of values in each column.

The result of this function is a data frame.

`read.fwf(file, widths)`

Read a text file in fixed-width format. The name of the file is specified by `file` and `widths` is a numeric vector specifying the width of each column of values.

The result is a data frame.

`read.csv(file)`

A front end for `read.table()` with default argument settings designed for reading a text file in CSV format.

The result is a data frame.

`read.delim(file)`

A front end for `read.table()` with default argument settings designed for reading a tab-delimited text file.

The result is a data frame.

`scan(file, what)`

Read data from a text file and produce a vector of values. The type of the value provided for the argument `what` determines how the values in the text file are interpreted. If this argument is a list, then the result is a list of vectors, each of a type corresponding to the relevant component of `what`.

This function is more flexible and faster than `read.table()` and its kin, but the result may be less convenient to work with.

In most cases, these functions that read a data set from a text file produce a data frame as the result. The functions automatically determine the data type for each column of the data frame, treating anything that is not a number as a factor, but arguments are provided to explicitly specify the data types for columns. Where names of columns are provided in the text file, these functions may modify the names so that they do not cause syntax problems, but again arguments are provided to stop these modifications from happening.

The **XML** package provides functions for reading and manipulating XML documents.

The package **foreign** contains various functions for reading data from external files in the various binary formats of popular statistical programs. Other popular scientific binary formats can also be read using an appropriate package, e.g., **ncdf** for the netCDF format.

Most of the functions for reading files have a corresponding function to write the relevant format.

`writeLines(text, con)`

Write a character vector to a text file. Each element of the character vector is written as a separate line in the file.

`write.table(x, file, sep=" ")`

Write a data frame to a text file using a delimited format. The `sep` argument allows control over the delimiter.

The function `write.csv()` provides useful defaults for producing files in CSV format.

`sink(file)`

Redirect R output to a text file. Instead of displaying output on the screen, output is saved into a file. The redirection is terminated by calling `sink()` with no arguments.

The function `capture.output()` provides a convenient way to redirect output for a single R expression.

Most of these functions read or write an entire file worth of data in one go. For large data sets, it is also possible to read or write data in smaller pieces. The functions `file()` and `close()` allow a file to be held open while reading or writing. Functions that read from files typically have an argument that specifies a number of lines or bytes of information to read, and functions that write to files typically provide an `append` argument to ensure that previous content is not overwritten.

One important case not mentioned so far is the export and import of data in an R-specific format, which is useful for sharing data between colleagues who all use R.

`save(..., file)`

Save the symbols named in `...` (and their values), in an R-specific format, to the specified `file`.

`load(file)`

Load R symbols (and their values) from the specified `file` (that has been created by a previous call to `save()`).

`dump(list, file)`

Write out a text representation of the R data structures named in the character vector `list`. The data structures can be recreated in R by calling `source()` on the `file`.

`source(file)`

Parse and evaluate the R code in `file`. This can be used to read data from a file created by `dump()` or much more generally to run any R code that has been stored in a file.

10.3.9 Transformations

`transform(data, ...)`

Redefine existing columns within a data frame and append new columns to a data frame.

Each argument in `...` is of the form *columnName=columnValue*.

`ifelse(test, yes, no)`

The `test` argument is a logical vector. This function creates a new vector consisting of the values in the vector `yes` when the corresponding element of `test` is `TRUE` and the values in `no` when `test` is `FALSE`.

The `switch()` function is similar, but allows for more than two values in `test`.

`cut(x, breaks)`

Transform the continuous vector `x` into a factor. The `breaks` argument can be an integer that specifies how many different levels to break `x` into, or it can be a vector of interval boundaries that are used to cut `x` into different levels.

An additional `labels` argument allows labels to be specified for the levels of the new factor.

10.3.10 Sorting

`sort(x, decreasing=FALSE)`

Sort the elements of a vector. Character values are sorted alphabetically (which may depend on the locale or language setting).

`order(..., decreasing=FALSE)`

Determine an ordering based on the elements of one or more vectors. In the simple case of a single vector, `sort(x)` is equivalent to `x[order(x)]`. The advantage of this function is that it can be used to reorder more than just a single vector, plus it can produce an ordering from more than one vector; it can break ties in one variable using the values from another variable.

10.3.11 Tables of counts

`table(...)`

Generate table of counts for one or more factors. The result is a `"table"` data structure, with as many dimensions as there are arguments.

The `margin.table()` function reduces a table to marginal totals, `prop.table()` converts table counts to proportions of marginal totals, and `addmargins()` *adds* margin totals to an existing table.

`xtabs(formula, data)`

Similar to `table()` except factors to cross-tabulate are expressed in a formula. Symbols in the formula will be searched for in the data frame given by the `data` argument.

`ftable(...)`
Similar to `table()` except that the result is always a two-dimensional `"ftable"` data structure, no matter how many factors are used. This makes for a more readable display.

10.3.12 Aggregation

`aggregate(x, by, FUN)`
Call the function `FUN` for each subset of `x` defined by the grouping factors in the list `by`. It is possible to apply the function to multiple variables (`x` can be a data frame) and it is possible to group by multiple factors (the list `by` can have more than one component). The result is a data frame. The names used in the `by` list are used for the relevant columns in the result. If `x` is a data frame, then the names of the variables in the data frame are used for the relevant columns in the result.

10.3.13 The "apply" functions

`apply(X, MARGIN, FUN, ...)`
Call a function on each row or each column of a data frame or matrix. The function `FUN` is called for each row of the matrix `X`, if `MARGIN=1`; if `MARGIN=2`, the function is called for each column of `X`. All other arguments are passed as arguments to `FUN`.

The data structure that is returned depends on the value returned by `FUN`. In the simplest case, where `FUN` returns a single value, the result is a vector with one value per row (or column) of the original matrix `X`.

`sweep(x, MARGIN, STATS, FUN="-")`
If `MARGIN=1`, for row i of `x`, subtract element i of `STATS`. For example, subtract row averages from all rows.

More generally, call the function `FUN` with row i of `x` as the first argument and element i of `STATS` as the second argument.

If `MARGIN=2`, call `FUN` for each column of `x` rather than for each row.

`tapply(X, INDEX, FUN, ...)`
Call a function once for each subset of the vector `X`, where the subsets correspond to unique values of the factor `INDEX`. The `INDEX` argument can be a list of factors, in which case the subsets are unique combinations of the levels of these factors.

The result depends on how many factors are given in `INDEX`. For the simple case where there is only one factor and `FUN` returns a single value, the result is a vector.

`lapply(X, FUN, ...)`
Call the function `FUN` once for each component of the list `X`. The result is a list. Additional arguments are passed on to each call to `FUN`.

`sapply(X, FUN, ...)`
Similar to `lapply()`, but will simplify the result to a vector if possible (e.g., if all components of `X` are vectors and `FUN` returns a single value).

`mapply(FUN, ..., MoreArgs)`
A "multivariate" apply. Similar to `lapply()`, but will call the function `FUN` on the first element of each of the supplied arguments, then on the second element of each argument, and so on. `MoreArgs` is a list of arguments to pass to each call to `FUN`.

`rapply(object, f)`
A "recursive" apply. Calls the function `f` on each component of the list `object`, *but* if a component is itself a list, then `f` is called on each component of that list as well, and so on.

10.3.14 Merging

`rbind(...)`
Create a new data frame by combining two or more data frames that have the same columns. The result is the union of the rows of the original data frames. This function also works for matrices.

`cbind(...)`
Create a new data frame by combining two or more data frames that have the same number of rows. The result is the union of the columns of the original data frames. This function also works for matrices.

`merge(x, y)`
Create a new data frame by combining two data frames in a database join operation. The two data frames will usually have different columns, though they will typically share at least one column, which is used to match the rows. Additional arguments allow the matching column to be specified explicitly.

The default join is an inner join on columns that `x` and `y` have in common. Additional arguments allow for the equivalent of inner joins and outer joins.

10.3.15 Splitting

`split(x, f)`

Split a vector or data frame, `x`, into a list of smaller vectors or data frames. The factor `f` is used to determine which elements of the original vector or which rows of the original matrix end up in each subset.

`unsplit(value, f)`

Combine a list of vectors into a single vector. The factor `f` determines the order in which the elements of the vectors are combined.

This function can also be used to combine a list of data frames into a single data frame (as long as the data frames have the same number of columns); in this case, `f` determines the order in which the rows of the data frames are combined.

10.3.16 Reshaping

`stack(x)`

Stack the existing columns of data frame `x` together into a single column and add a new column that identifies which original column each value came from.

`aperm(a, perm)`

Reorder the dimensions of an array. The `perm` argument specifies the order of the dimensions.

The special case of transposing a matrix is provided by the `t()` function.

Functions from the **reshape** package:

`melt(data, measure.var)`
`melt(data, id.var)`

Convert the `data`, typically a data frame, into "long" form, where there is a row for every measurement or "dependent" value. The `measure.var` argument gives the names or numeric indices of the variables that contain measurements. All other variables are treated as labels characterizing the measurements (typically factors). Alternatively, the `id.var` argument specifies the label variables and all others are treated as measurements.

In the resulting data frame, there is a new, single column of measurements with the name `value` and an additional variable of identifying labels, named `variable`.

`cast(data, formula)`

Given `data` in a long form, i.e., produced by `melt()`, restructure the data according to the given `formula`. In the new arrangement, variables mentioned on the left-hand side of the formula vary across rows and variables mentioned on the right-hand side vary across columns.

In a simple repeated-measures scenario consisting of measurements at two time points, the data may consist of a variable of subject IDs plus two variables containing measurements at the two time points.

```
R> library(reshape)
R> wide <- data.frame(ID=1:3,
                      T1=rnorm(3),
                      T2=sample(100:200, 3))
R> wide
```

```
  ID          T1  T2
1  1 -0.07942034 167
2  2  0.58249374 157
3  3 -0.49222018 151
```

If we melt the data, we produce a data frame with a column named ID, a column named `variable` with values `T1` or `T2`, and a column named `value`, containing all of the measurements.

```
R> long <- melt(wide,
                id.var=c("ID"),
                measure.var=c("T1", "T2"))
R> long
```

```
  ID variable        value
1  1       T1  -0.07942034
2  2       T1   0.58249374
3  3       T1  -0.49222018
4  1       T2 167.00000000
5  2       T2 157.00000000
6  3       T2 151.00000000
```

This form can be recast back to the original wide form as follows.

```
R> cast(long, ID ~ variable)

  ID          T1  T2
1  1 -0.07942034 167
2  2  0.58249374 157
3  3 -0.49222018 151
```

The function `recast()` combines a melt and cast in a single operation.

10.3.17 Text processing

`nchar(x)`
: Count the number of characters in each element of the character vector `x`. The result is a numeric vector the same length as `x`.

`grep(pattern, x)`
: Search for the regular expression `pattern` in the character vector `x`. The result is a numeric vector identifying which elements of `x` matched the pattern. If there are no matches, the result has length zero.

  The function `agrep()` allows for approximate matching.

`regexpr(pattern, text)`
: Similar to `grep()` except that the result is a numeric vector containing the character location of the match within each element of `text` (`-1` if there is no match). The result also has an attribute, `match.length`, containing the length of the match.

`gregexpr(pattern, text)`
: Similar to `regexpr()`, except that the result is the locations (and lengths) of *all* matches within each piece of text. The result is a *list*.

`gsub(pattern, replacement, x)`
: Search for the regular expression `pattern` in the character vector `x` and replace all matches with the character value in `replacement`. The result is a vector containing the modified text.

  The `g` stands for "global" so all matches are replaced; there is a `sub()` function that just replaces the first match.

  The functions `toupper()` and `tolower()` convert character values to all uppercase or all lowercase.

`substr(x, start, stop)`
: For each character value in `x`, return a subset of the text consisting of

the characters at positions `start` through `stop`, inclusive. The first character is at position `1`.

The function `substring()` works very similarly, with the extra convenience that the end character defaults to the end of the text.

More specialized text subsetting is provided by the `strtim()` function, which removes characters from the end of text to enforce a maximum length, and `abbreviate()`, which reduces text to a given length by removing characters in such a way that each piece of text remains unique.

`strsplit(x, split)`
: For each character value in `x`, break the text into separate character values, using `split` as the delimiter. The result is a *list*, with one character vector component for each element of the original vector `x`.

`paste(..., sep, collapse)`
: Combine text, placing the character value `sep` in between. The result is a character vector the same length as the *longest* of the arguments, so shorter arguments are recycled. If the `collapse` argument is not `NULL`, the result vector is then collapsed to a single character value, with the text `collapse` placed in between each element of the result.

10.3.18 Data display

`print(x)`
: This function generates most of the output that we see on the screen. The important thing to note is that this function is generic, so the output will depend on the class of `x`. For different classes there are also different arguments to control the way the output is formatted. For example, there is a `digits` argument that controls the number of significant digits that are printed for numeric values, and there is a `quote` argument that controls whether double-quotes are printed around character values.

`format(x)`
: The usefulness of this function is to produce a character representation of a data structure where all values have a similar format; for example, all numeric values are formatted to have the same number of characters in total.

`sprintf(fmt, ...)`
: Generates a character vector using the template given by `fmt`. This is a character value with special codes embedded. Each special code

provides a placeholder for values supplied as further arguments to the `sprintf()` function and the special code controls the formatting of the values. The help page for this function contains a description of the special codes and their meanings.

The usefulness of this function is to obtain fine control over formatting that is not available in `print()` or `format()`.

`strwrap(x, width)`
: Break long pieces of text, by inserting newlines, so that each line is less than the given `width`.

`cat(..., sep=" ", fill=FALSE)`
: Displays the values in `...` on screen. This function converts its arguments to character vectors if necessary and performs no additional formatting of the text it is given.

  The `fill` argument can be used to control when to start a new line of output, and the `sep` argument specifies text to place between arguments.

10.3.19 Debugging

The functions from the previous section are useful to display intermediate results from within a loop or function.

`debug(fun)`
: Following a call to this function, the function `fun` will be run one expression at a time, rather than all at once. After each expression, the values of symbols used within the function can be explored by typing the symbol name. Typing 'n' (or just hitting Enter) runs the next expression; 'c' runs all remaining expressions; and 'Q' quits from the function.

10.4 Getting help

The `help()` function is special in that it provides information about other functions. This function displays a **help page**, which is online documentation that describes what a function does. This includes an explanation of all of the arguments to the function and a description of the result produced by the function. Figure 10.1 shows the beginning of the help page for the `Sys.sleep()` function, which is obtained by typing `help(Sys.sleep)`.

```
Sys.sleep                  package:base              R Documentation

Suspend Execution for a Time Interval

Description:

     Suspend execution of R expressions for a given number of
     seconds

Usage:

     Sys.sleep(time)

Arguments:

    time: The time interval to suspend execution for, in seconds.

Details:

     Using this function allows R to be given very low priority
     and hence not to interfere with more important foreground
     tasks. A typical use is to allow a process launched from R
     to set itself up and read its input files before R execution
     is resumed.
```

Figure 10.1: The beginning of the help page for the function `Sys.sleep()` as displayed within an xterm on a Linux system. This help page is displayed by the expression `help(Sys.sleep)`.

A special shorthand using the question mark character, ?, is provided for getting the help page for a function. Instead of typing help(Sys.sleep), it is also possible to simply type ?Sys.sleep.

There is also a web interface to the help system, which is initiated by typing help.start().

Many help pages also have a set of examples to demonstrate the proper use of the function, and these examples can be run using the example() function.

10.5 Packages

R functions are organized into collections of functions called **packages**. A number of packages are installed with R by default and several packages are loaded automatically in every R session. The search() function shows which packages are currently available, as shown below:

```
R> search()
```

```
[1] ".GlobalEnv"          "package:stats"       "package:graphics"
[4] "package:grDevices"   "package:utils"       "package:datasets"
[7] "package:methods"     "Autoloads"           "package:base"
```

The top line of the help page for a function shows which package the function comes from. For example, Sys.sleep() comes from the **base** package (see Figure 10.1).

Other packages may be loaded using the library() function. For example, the **foreign** package provides functions for reading in data sets that have been stored in the native format of a different statistical software system. In order to use the read.spss() function from this package, the **foreign** package must be **loaded** as follows:

```
R> library(foreign)
```

The search() function confirms that the **foreign** package is now loaded and all of the functions from that package are now available.

```
R> search()
```

```
 [1] ".GlobalEnv"        "package:foreign"   "package:stats"
 [4] "package:graphics"  "package:grDevices" "package:utils"
 [7] "package:datasets"  "package:methods"   "Autoloads"
[10] "package:base"
```

There are usually 25 packages distributed with R. Over a thousand other packages are available for download from the web via the Comprehensive R Archive Network (CRAN), `http://cran.r-project.org`. These packages must first be **installed** before they can be loaded. A new package can be installed using the `install.packages()` function (on Windows, there is an option on the Packages menu).

10.6 Searching for functions

Given the name of a function, it is not difficult to find out what that function does and how to use the function by reading the function's help page. A more difficult job is to find the name of a function that will perform a particular task.

The `help.search()` function can be used to search for functions relating to a keyword within the current R installation. The `RSiteSearch()` function performs a more powerful and comprehensive web-based search of functions in almost all known R packages, R mailing list archives, and the main R manuals. This is based on Jonathan Baron's search site, `http://finzi.psych.upenn.edu/search.html`. There is also a Google customized search available, `http://www.rseek.org`, which provides a convenient categorization of the search results. This was set up and is maintained by Sasha Goodman.

Another problem that arises is that, while information on a single function is easy to obtain, it can be harder to discover how several related functions work together. One way to get a broader overview of functions in a package is to read a package **vignette** (see the `vignette()` function). There are also overviews of certain areas of research or application provided by CRAN Task Views (one of the main links on the CRAN web site) and there is a growing list of books on R (see the home page of the R Project).

10.7 Further reading

The Home Page of the R Project

`http://www.r-project.org/`

The starting point for finding information about R. Includes manuals, particularly "An Introduction to R" and "R Data Import/Export", and FAQ's.

CRAN—the Comprehensive R Archive Network

`http://cran.r-project.org/`

Where to download R and the various add-on packages for R.

R News

`http://cran.r-project.org/doc/Rnews/`

The newsletter of the R Project. Issue 4(1) has an excellent article on working with dates in R.

Introductory Statistics with R

by Peter Dalgaard

2nd Edition (2008) Springer.

Introduction to using R for basic statistical analysis.

Data Manipulation with R

by Phil Spector

(2008) Springer.

Broad coverage of data manipulation functions in R.

R Coding Conventions

`http://www1.maths.lth.se/help/R/RCC/`

An unofficial set of guidelines for laying out R code and choosing symbol names for R code.

11

Regular Expressions Reference

Regular expressions provide a very powerful way to describe general patterns in text.

This chapter only provides a quick reference for regular expressions. See Section 9.9.2 for an explanation of what regular expressions are, what they are useful for, and for more realistic examples of their usage.

Small examples are provided throughout this chapter in order to demonstrate the meaning of various features of regular expressions. In each case, a regular expression will be used to search for a pattern within the following text:

```
The cat sat on the mat.
```

A successful match will be indicated by highlighting the matched letters within the text. For example, if the pattern being searched for is the word `"on"`, the the result will be displayed as shown below, with the pattern on the left and the resulting match(es) on the right.

`"on"` ⇒ `The cat sat on the mat.`

11.1 Literals

In a regular expression, any character that is not a metacharacter has its normal literal meaning. This includes the lowercase and uppercase alphabet and the digits, plus some punctuation marks, such as the colon and the exclamation mark. Many other punctuation characters and symbols are metacharacters and have a special meaning.

For example, the regular expression `"cat"` means: the letter 'c', followed by the letter 'a', followed by the letter 't'.

`"cat"` ⇒ `The cat sat on the mat.`

11.2 Metacharacters

The simplest example of a metacharacter is the full stop.

'.'

The full stop character matches any single character of any sort (apart from a newline).

For example, the regular expression ".at" means: *any* letter, followed by the letter 'a', followed by the letter 't'.

".at" ⇒ The <u>cat</u> <u>sat</u> on the <u>mat</u>.

11.2.1 Character sets

'[' and ']'

Square brackets in a regular expression are used to indicate a character set. A character set will match any character in the set.

For example, the regular expression "[Tt]he" means: an uppercase 'T' *or* a lowercase 't', followed by the letter 'h', followed by the letter 'e'.

"[Tt]he" ⇒ <u>The</u> cat sat on <u>the</u> mat.

Within square brackets, common ranges may be specified by start and end characters, with a dash in between, e.g., 0-9 for the Arabic numerals.

For example, the regular expression "[a-z]at" means: any (English) lowercase letter, followed by the letter 'a', followed by the letter 't'.

"[a-z]at" ⇒ The <u>cat</u> <u>sat</u> on the <u>mat</u>.

If a hat character appears as the first character within square brackets, the set is inverted so that a match occurs if any character other than the set specified within the square brackets is found.

For example, the regular expression "[^c]at" means: any letter *except* 'c', followed by the letter 'a', followed by the letter 't'.

"[^c]at" ⇒ The cat <u>sat</u> on the <u>mat</u>.

Table 11.1: Some POSIX regular expression character classes.

`[:alpha:]`	Alphabetic (only letters)
`[:lower:]`	Lowercase letters
`[:upper:]`	Uppercase letters
`[:digit:]`	Digits
`[:alnum:]`	Alphanumeric (letters and digits)
`[:space:]`	White space
`[:punct:]`	Punctuation

Within square brackets, most metacharacters revert to their literal meaning. For example, `[.]` means a literal full stop.

For example, the regular expression `"at[.]"` means: the letter 'a', followed by the letter 't', followed by a full stop.

`"at."` ⇒ `The cat sat on the mat.`

`"at[.]"` ⇒ `The cat sat on the mat.`

In POSIX regular expressions, common character ranges can be specified using special character sequences of the form `[:`*keyword*`:]` (see Table 11.1). The advantage of this approach is that the regular expression will work in different languages. For example, `[a-z]` will not capture all characters in languages that include accented characters, but `[[:alpha:]]` will.

For example, the regular expression `"[[:lower:]]at"` means: any lowercase letter in any language, followed by the letter 'a', followed by the letter 't'.

`"[[:lower:]]at"` ⇒ `The cat sat on the mat.`

11.2.2 Anchors

Anchors do not match characters. Instead, they match zero-length features of a piece of text, such as the start and end of the text.

`'^'`

The "hat" character matches the start of a piece of text or the start of a line of text. Putting this at the start of a regular expression forces the match to begin at the start of the text. Otherwise, the pattern can be matched anywhere within the text.

For example, the regular expression `"^[Tt]he"` means: *at the start of the text* an uppercase 'T' *or* a lowercase 't', followed by the letter 'h', followed by the letter 'e'.

`"[Tt]he"` ⇒ `The cat sat on the mat.`

`"^[Tt]he"` ⇒ `The cat sat on the mat.`

'$'

The dollar character matches the end of a piece of text. This is useful for ensuring that a match finishes at the end of the text.

For example, the regular expression `"at.$"` means: the letter 'a', followed by the letter 't', followed by any character, *at the end of the text.*

`"at."` ⇒ `The cat sat on the mat.`

`"at.$"` ⇒ `The cat sat on the mat.`

11.2.3 Alternation

'|'

The vertical bar character subdivides a regular expression into alternative subpatterns. A match is made if either the pattern to the left of the vertical bar or the pattern to the right of the vertical bar is found.

For example, the regular expression `"cat|sat"` means: the letter 'c', followed by the letter 'a', followed by the letter 't', *or* the letter 's', followed by the letter 'a', followed by the letter 't'.

`"cat|sat"` ⇒ `The cat sat on the mat.`

Pattern alternatives can be made a subpattern within a large regular expression by enclosing the vertical bar and the alternatives within parentheses (see Section 11.2.5 below).

11.2.4 Repetitions

Three metacharacters are used to specify how many times a subpattern can repeat. The repetition relates to the subpattern that immediately precedes the metacharacter in the regular expression. By default, this is just the previous character, but if the preceding character is a closing square bracket or a closing parenthesis then the modifier relates to the entire character set or the entire subpattern within the parentheses (see Section 11.2.5 below).

'?'

The question mark means that the subpattern can be missing or it can occur exactly once.

For example, the regular expression "at[.]?" means: the letter 'a', followed by the letter 't', *optionally* followed by a full stop.

"at[.]" ⇒ The cat sat on the mat.

"at[.]?" ⇒ The cat sat on the mat.

'*'

The asterisk character means that the subpattern can occur *zero* or more times.

'+'

The plus character means that the subpattern can occur *one* or more times.

For example, the regular expression "[a-z]+" means: any number of (English) lowercase letters in a row.

"[a-z]+" ⇒ The cat sat on the mat.

As an example of how regular expressions are "greedy", the regular expression "c.+t" means: the letter 'c', followed by any number of any character at all, followed by the letter 't'.

"c.+t" ⇒ The cat sat on the mat.

11.2.5 Grouping

'(' and ')'

Parentheses can be used to define a subpattern within a regular expression. This is useful for providing alternation just within a subpattern of a regular expression and for applying a repetition metacharacter to more than a single character.

For example, the regular expression `"(c|s)at"` means: the letter 'c' *or* the letter 's', followed by the letter 'a', followed by the letter 't'.

`"(c|s)at"` ⇒ The cat sat on the mat.

The regular expression `"(.at )+"` means *one or more repetitions of* the following pattern: any letter at all, followed by the letter 'a', followed by the letter 't', followed by a space.

`"(.at )+"` ⇒ The cat sat on the mat.

Grouping is also useful for retaining original portions of text when performing a search-and-replace operation (see Section 11.2.6).

11.2.6 Backreferences

It is possible to refer to previous subpatterns within a regular expression using **backreferences**.

'\n'

When parentheses are used in a pattern to delimit subpatterns, each subpattern may be referred to using a special escape sequence: `\1` refers to the first subpattern (reading from the left), `\2` refers to the second subpattern, and so on.

For example, the regular expression `"c(..) s\\1"` means: the letter 'c', followed by *any two letters*, followed by a space, followed by the letter 's', followed by *whichever two letters followed the letter 'c'*.

`"c(..) s\\1"` ⇒ The cat sat on the mat.

When performing a search-and-replace operation, backreferences may be used to specify that the text matched by a subpattern should be used as part of the replacement text.

For example, in the first replacement below, the literal text 'cat' is replaced with the literal text 'cot', but in the second example, any three-letter word ending in 'at' is replaced by a three-letter word *with the original starting letter* but ending in 'ot'.

```
R> gsub("cat", "cot", text)

[1] "The cot sat on the mat."
```

```
R> gsub("(.)at", "\\1ot", text)

[1] "The cot sot on the mot."
```

Notice that, within an R expression, the backslash character must be escaped as usual, so the replacement text referring to the first subpattern would have to written like this: `"\\1"`.

Some more realistic examples of the use of backreferences are given in Section 9.9.3.

11.3 Further reading

Mastering Regular Expressions
by Jeffrey Friedl
3rd Edition (2006) O'Reilly.
Exhaustive treatment of regular expressions in all their various incarnations.

Regular Expression Pocket Reference
by Tony Stubblebine
2nd Edition (2007) O'Reilly.
Quick reference for regular expressions.

Regular-Expressions.info
`http://wwww.regular-expressions.info`
Comprehensive online resource.

12 Conclusion

This book has described how a number of important data technologies work *and* it has provided guidelines on the right way to use those technologies.

We have covered technologies for storing data, accessing data, processing data, and publishing data on the web. We have also focused on writing computer languages.

We have looked at the advantages and disadvantages of different storage options, we have looked at how to design data storage efficiently, and we have discussed how to write *good* computer code.

There are a number of important ideas to take away from this book.

One of these is that the computer is a very flexible and powerful tool, and it is a tool that is ours to control. Files and documents, especially those in open standard formats, can be manipulated using a variety of software tools, not just one specific piece of software. A programming language is a tool that allows us to manipulate data stored in files and to manipulate data held in RAM in unlimited ways. Even with a basic knowledge of programming, we can perform a huge variety of data processing tasks.

A related idea is that computer code is the preferred approach to communicating our instructions to the computer. This approach allows us to be precise and expressive, it provides a complete record of our actions, and it allows others to replicate our work.

Writing computer code to process, store, or display data is a task that should be performed with considerable discipline. It is important to develop code in small pieces and in careful stages, and it is important to produce code that is tidy and sensibly structured. This discipline is essential to writing code that will produce the correct result both now and in the future.

Another important idea is the DRY principle. Information, whether data or computer code, should be organized in such a way that there is only one copy of each important unit of information. We saw this idea most clearly in terms of data storage, in the design of XML documents and in the design of relational databases. But the ideas of organizing information efficiently can influence how we work in many ways. The DRY principle can also be applied to how we write computer code, particularly in the use of loops and functions. It can also be applied to how we collect our code within files and how we organize files within directories or folders.

The aim of this book is to expose and demystify some of the details and underlying principles of how data technologies work, so that we can unlock the power and potential of the "universal computing machines" that sit on our desktops.

Attributions

This book makes use of a number of images that are in the public domain or that have been released under permissive licenses that encourage sharing and reuse. The authors and sources of these images are acknowledged below.

 Section 1.1.
Author: Particle Dynamics Group, Department of Oceanography, Texas A&M University.
Source: http://www-ocean.tamu.edu/~pdgroup/jpegs/waves.jpg
License: Used and distributed with permission.

 Section 5.1.
Author: Bill Casselman.
Source: http://upload.wikimedia.org/wikipedia/commons/0/0b/Ybc7289-bw.jpg
License: Creative Commons Attribution 2.5.

 Section 5.2.8.
Author: T3rminatr
Source: http://commons.wikimedia.org/wiki/Image:Orbital_Planes.svg
License: Public Domain.

 Section 5.4.3.
Author: Klaus Gena.
Source: http://openclipart.org/people/KlausGena/KlausGena_Tuned_Lada_VAZ_2101.svg
License: Public Domain.

 Section 7.2.5.
Author: Denelson83 http://en.wikipedia.org/wiki/User:Denelson83.
Source: http://commons.wikimedia.org/wiki/Image:Flag_of_the_Commonwealth_of_Nations.svg
License: Public Domain.

 Section 9.1.
Author: Nukkio http://www.nukkio.altervista.org/index.htm.
Source: http://openclipart.org/people/ernes/ernes_orologio_clock.svg
License: Public Domain.

 Section 9.8.12.
Author: Balder777 http://de.wikipedia.org/wiki/Benutzer:Balder777.
Source: http://commons.wikimedia.org/wiki/Image:Spiralförmige_Energiesparlampe_quadr.png
License: Public Domain.

 Section 9.9.3.
Author: United States Department of Agriculture.
Source: http://commons.wikimedia.org/wiki/Image:Usda_rachis_barley.jpg
License: Public Domain.

Bibliography

[1] Daniel Adler, Christian Gläser, Oleg Nenadic, Jens Oehlschlägel, and Walter Zucchini. *ff: memory-efficient storage of large atomic vectors and arrays on disk and fast access functions*, 2008. R package version 2.0.0.

[2] Henrik Bengtsson. *R Coding Conventions.* `http://www1.maths.lth.se/help/R/RCC/`, 2008.

[3] Tim Berners-Lee. *The World Wide Web: Past, Present and Future*, 1996. `http://www.w3.org/People/Berners-Lee/1996/ppf.html`.

[4] Tim Berners-Lee. *The World Wide Web: A Very Short Personal History*, 2007. `http://www.w3.org/People/Berners-Lee/ShortHistory.html`.

[5] Rafe Colburn. *Using SQL (Special Edition).* Que, 1999.

[6] David B. Dahl. *xtable: Export tables to LaTeX or HTML*, 2007. R package version 1.5-2.

[7] Peter Dalgaard. *Introductory Statistics with R.* Springer, 2nd edition, 2008.

[8] Data Documentation Initiative. *DDI home page*, 2008. `http://www.ddialliance.org/`.

[9] DevShed. *DevShed home page*, 2008. `http://www.devshed.com/`.

[10] FileFormat.Info. *FileFormat.Info: The Digital Rosetta Stone*, 2008. `http://www.fileformat.info/`.

[11] Firefox home page. `http://www.mozilla.com/en-US/products/firefox/`, 2008.

[12] Jeffrey Friedl. *Mastering Regular Expressions.* O'Reilly, 3rd edition, 2006.

[13] Jan Goyvaerts. Regular-Expressions.info home page. `http://wwww.regular-expressions.info`, 2008.

[14] James R. Groff and Paul N. Weinberg. *SQL: The Complete Reference.* McGraw-Hill Osborne Media, 2nd edition, 2002.

[15] Gabor Grothendieck and Thomas Petzoldt. R help desk: Date and time classes in R. *R News*, 4(1):29–32, 2004.

[16] Jeffrey A. Hoffer, Mary Prescott, and Fred McFadden. *Modern Database Management.* Prentice Hall, 7th edition, 2004.

[17] Andrew Hunt and David Thomas. *The Pragmatic Programmer: From Journeyman to Master.* Addison-Wesley, 1999.

[18] IBM. *developerWorks home page*, 2008. `http://www.ibm.com/developerworks/`.

[19] David A. James. *RSQLite: SQLite interface for R*, 2008. R package version 0.6-3.

[20] David A. James and Saikat DebRoy. *RMySQL: R interface to the MySQL database*, 2006. R package version 0.5-11.

[21] Michael Lapsley and B. D. Ripley. *RODBC: ODBC Database Access*, 2008. R package version 1.2-3.

[22] Friedrich Leisch. Sweave: Dynamic generation of statistical reports using literate data analysis. In Wolfgang Härdle and Bernd Rönz, editors, *Compstat 2002 — Proceedings in Computational Statistics*, pages 575–580. Physica Verlag, Heidelberg, 2002.

[23] Håkon Wium Lie and Bert Bos. *Cascading Style Sheets, Level 1.* World Wide Web Consortium (W3C), 1996. `http://www.w3.org/TR/CSS1/`.

[24] Steve McConnell. *Code Complete: A Practical Handbook of Software Construction.* Microsoft Press, 2nd edition, 2004.

[25] Paul Murrell. *hexView: Viewing Binary Files*, 2008. R package version 0.4.

[26] Gregoire Pau. *hwriter: HTML Writer*, 2008. R package version 0.92.

[27] Roger D. Peng. Interacting with data using the filehash package. *R News*, 6(4):19–24, 2006.

[28] David Pierce. *ncdf: Interface to Unidata netCDF data files*, 2008. R package version 1.6.

[29] R Development Core Team. *R: A Language and Environment for Statistical Computing.* R Foundation for Statistical Computing, Vienna, Austria, 2008.

[30] R Foundation for Statistical Computing. *Comprehensive R Archive Network*, 2008. `http://cran.r-project.org/`.

[31] R Foundation for Statistical Computing. *R Project for Statistical Computing home page*, 2008. `http://www.r-project.org/`.

[32] R Special Interest Group on Databases (R-SIG-DB). *DBI: R Database Interface*, 2008. R package version 0.2-3.

[33] Dave Raggett. Getting started with HTML. `http://www.w3.org/MarkUp/Guide/`, 2005.

[34] Dave Raggett. Adding a touch of style. `http://www.w3.org/MarkUp/Guide/Style.html`, 2008.

[35] Dave Raggett. HTML Tidy home page. `http://www.w3.org/People/Raggett/tidy/`, 2008.

[36] Dave Raggett, Jenny Lam, Ian F. Alexander, and Michael Kmiec. *Raggett on HTML 4*. Addison-Wesley, 2nd edition, 1997.

[37] Roman Czyborra. *The ISO 8859 Alphabet Soup*, 1998. `http://aspell.net/charsets/iso8859.html`.

[38] Dave Shea. Zen Garden home page. `http://www.csszengarden.com/`, 2008.

[39] Phil Spector. *Data Manipulation with R*. Springer, 2008.

[40] SQLite home page. `http://www.sqlite.org/`, 2008.

[41] Statistical Data and Metadata Exchange Initiative. *SDMX User Guide*, 2007. `http://sdmx.org/wp-content/uploads/2008/02/sdmx-userguide-version2007-1-4.doc`.

[42] Tony Stubblebine. *Regular Expression Pocket Reference*. O'Reilly Media, Inc., 2nd edition, 2007.

[43] Hans-Peter Suter. *xlsReadWrite: Natively read and write Excel files*, 2008. R package version 1.3.2.

[44] Duncan Temple Lang. *XML: Tools for parsing and generating XML within R and S-Plus*, 2008. R package version 1.95-3.

[45] U.S. Census Bureau. *World POPClock Projection*, 2008. `http://www.census.gov/ipc/www/popclockworld.html`.

[46] Daniel Veillard. libxml home page. `http://xmlsoft.org/`, 2008.

[47] W3Schools. CSS tutorial. `http://www.w3schools.com/css/`, 2008.

[48] W3Schools. DTD tutorial. `http://www.w3schools.com/dtd/`, 2008.

[49] W3Schools. HTML tutorial. `http://www.w3schools.com/html/`, 2008.

[50] W3Schools. XML tutorial. `http://www.w3schools.com/xml/`, 2008.

[51] W3Schools. XPath tutorial. `http://www.w3schools.com/Xpath/`, 2008.

[52] Gregory R. Warnes. *gdata: Various R programming tools for data manipulation*, 2005. R package version 2.1.2.

[53] Web Design Group. *Cascading Style Sheets*, 2008. `http://htmlhelp.com/reference/css/`.

[54] Web Design Group. *HTML 4 Reference*, 2008. `http://htmlhelp.com/reference/html40/`.

[55] Hadley Wickham. *reshape: Flexibly reshape data*, 2007. R package version 0.8.0.

[56] Wikipedia. Pole of inaccessibility, 2008. [Online; accessed 22-September-2008].

[57] World Wide Web Consortium (W3C). *HTML 4.01 Specification*, 1999. `http://www.w3.org/TR/html401/`.

[58] World Wide Web Consortium (W3C). *Extensible Markup Language (XML) 1.0*, 4th edition, 2006. `http://www.w3.org/TR/2006/REC-xml-20060816/`.

[59] World Wide Web Consortium (W3C). CSS Validation Service home page. `http://jigsaw.w3.org/css-validator/`, 2008.

[60] World Wide Web Consortium (W3C). Markup validation service home page. `http://validator.w3.org/`, 2008.

[61] Achim Zeileis and Gabor Grothendieck. zoo: S3 infrastructure for regular and irregular time series. *Journal of Statistical Software*, 14(6):1–27, 2005.

Index